Using the Baofeng® Radio

Your Guide to Using a Handheld Ham Radio

by John Leonardelli, VE3IPS

START HERE

Your handheld radio is your gateway to communications via the Amateur Radio Service, also known as **ham radio**. Amateur radio is a federally regulated and licensed radio service for people of all ages. Ham radio provides the broadest and most powerful wireless communications capability available to any private citizen anywhere in the world.

To operate your radio, follow these essential steps to earn your **Amateur Radio License** first.

☐ **Pass your Amateur Radio License exam | www.arrl.org/shop**
Begin studying for your exam with these trusted resources:

The ARRL Ham Radio License Manual

Gordon West Technician Class Prep Book

☐ **Find an Exam Session | www.arrl.org/find-an-exam**
Once you're ready, locate a testing session near you to take the exam.

☐ **Join a Radio Club | www.arrl.org/find-a-club**
Being part of a local radio club is a great way to learn, network, and get support as you get on the air with your radio.

☐ **Volunteer in the Amateur Radio Emergency Service® — ARES® | www.arrl.org/ares**
Some hams train in groups that serve their communities during emergencies or other disasters. Learn how you can contribute your skills in times of need by joining ARES.

ARRL The National Association for Amateur Radio®
225 Main Street, Newington, CT 06111-1400 USA
www.arrl.org

Printed in the USA.

ISBN: 978-1-62595-224-0

First Edition
Second Printing

We strive to produce books without errors. Sometimes mistakes do occur, however. When we become aware of problems in our books (other than obvious typographical errors), we post corrections on the ARRL website.

If you think you have found an error, please check **www.arrl.org/product-notes** for corrections. If you don't find a correction there, please let us know by sending an email to **pubsfdbk@arrl.org**.

Contents

1 The Radio Services

1.1 Ham Radio

The Baofeng UV-5R is, in my opinion, an iconic and accessible handheld transceiver. An entry-level device, this compact radio offers versatile communication capabilities. Its affordability, dual band functionality (it operates on the 2 meter and 70 centimeter bands), and ease of use has made it an appealing option for those new to two-way radio and those who have been chatting on the airwaves for decades.

I originally purchased a UV-5R back in 2015 as a dedicated radio for an APRS project. I was so happy with how well it performed that I purchased a few more for go kits. Eventually, I started an informal breakfast series called the Toronto Baofeng Users Group (T-BUG) and started sharing information and code plugs.

For a fraction of the cost of higher-end radios, it delivers essential features such as adjustable power output, programmable memory channels, and FM broadcast reception. This makes it a solid beginner's choice for ham radio enthusiasts. For such a feature-rich radio, the UV-5R is easy to use. Its basic functionality isn't overwhelming for those who are just getting started with radio. For example, it's easy to manually program, and users can add numerous channels quickly with a programming cable and software (like *CHIRP*). Chapter 6 will lead you through all the user basics, and Chapter 7 will provide details on using *CHIRP* software. Many amateur radio enthusiasts keep a UV-5R as a backup or emergency radio because of its simplicity and ruggedness. Its affordable price makes it easy to store in a go-bag, car, or emergency kit without concern.

For someone starting in amateur radio, or needing an inexpensive, functional, and reliable radio, the UV-5R is a practical choice. However, the brand has expanded over the years with new models that offer higher power output and digital radio modes that appeal to a wider variety of users. To give a full picture of Baofeng radios' capabilities, this book also looks at the larger UV-82 and the DMR capable DR-1801UV radio.

Figure 1.1 — From left to right, the Baofeng DR-1801UV, UV-5R, and UV-82.

1.1.1 What is Ham Radio?

Amateur radio, often referred to as "ham radio," is a hobby and communication service that offers a fascinating way to communicate and explore the world of radio waves. Licensed operators use designated radio frequencies (allocated in band plans) for personal communications, during emergencies and technical experimentation. Ham radio is often referred to as the first social network because amateur radio enthusiasts built long distance communications networks with it long before the internet. It's a versatile and community-driven activity, involving a broad range of communication types, frequencies, and technologies. Ham radio fosters a sense of community locally through clubs and globally through on-air "ragchews" (informal chats). Operators can join conversations on repeaters, join nets (organized gatherings on-air), or participate in global networks that link repeaters and hotspots. Whether you're interested in personal conversations over the airwaves, emergency communications, building electronics, or simply chatting with fellow hams, ham radio has something for you.

Amateur radio will surprise you with all its different activities. If you've encountered amateur radio in a public service role or if someone you know has a ham radio in their home or car, then you already have some ideas of what's possible. Maybe you've seen ham radio in a movie or read about it in a book. Are you part of a maker community? If so, you'll really enjoy getting involved with one of the most hands-on hobbies of all. Amateur radio is the most powerful communications service available to the private citizen anywhere on earth—or even above it!

Figure 1.2 — Hams enjoy learning how their equipment works, as well as the satisfaction that comes from building their own equipment. This Morse code transceiver was assembled from a kit that included all of the parts, a circuit board, and detailed instructions.

Amateur radio is a recognized national asset, providing trained operators, technical specialists and disaster response communications in time of need. It was created for people just like you who have an interest in radio communications. Some hams prefer to focus on the technology and science of radio. Competitive events and awards programs hold the interest of others. some train to use radio in support of emergency response efforts, to provide public service, or to keep in touch with family. There are many hams who simply like to talk with other hams, too. This chapter, and the next several, will give you ideas for things you can do with your new Baofeng radio and an amateur radio license. Today, getting started in amateur radio is easier than ever due to accessible licensing options and a variety of user-friendly equipment. Whether you achieve the beginner or pass all three exams to get the most advanced license, the Baofeng radios mentioned here are cost-effective first options that will help you hit the airwaves and connect with fellow hams. ARRL, short for the American Radio Relay League, is National Association for Amateur Radio® in the United States. They can help you find amateur radio license preparation classes in your community and get you started in this wonderful hobby.

Figure 1.3 — Hams enjoy bringing portable radios to state and national parks to operate in a program called Parks on the Air (POTA).

Here's to embarking on your amateur radio adventure! Welcome to ham radio.

1.1.2 How to Obtain a Radio License

To legally operate a ham radio, you must have a license issued by a government authority. Licensing requires passing an exam that covers radio theory, operating practices, and regulations. There are currently three license levels available, with higher levels granting access to more frequencies and privileges. Licensed operators are issued call signs, which are used in communication with others and for identification. When I was licensed in 1977 as a basic amateur, I was issued the call sign VE3IPS. When I passed my advanced exam, I was able to also obtain the call sign VE3IQ.

In the United States, the Federal Communications Commission (FCC) governs amateur radio licensing. The three levels of amateur radio license are the Technician class, General class, and Amateur Extra class. Let's talk a bit about the differences between the different license classes.

License Classes

The Technician class is the starting point for new hams. It requires passing a written exam that covers basic radio theory, regulations, and operating procedures. Technicians can operate on VHF and UHF bands, which are ideal for local communications, and they can even access some HF bands for limited long-distance communication. The Technician exam consists of 35 multiple-choice questions, with the passing rate being around 80%. Many find the exam relatively straightforward after a few weeks of study, especially with the amount of studying material available online and in print. The exams are administered by Volunteer Examiner Coordinators (VECs). National VECs—like the one run by the ARRL—along with a variety of regional VECs make it easy to find a testing opportunity no matter where you're located.

Many people enjoy the hobby at this level, particularly those interested in VHF, UHF, and SHF frequencies, as well as high-speed data.

After obtaining your Technician license, you can pursue the General class license. This requires passing another written exam with more technical content. General class licensees gain access to the HF bands, allowing for global communications. The General exam also consists of 35 questions, which cover more complex topics, so you'll want to dedicate additional study time to the exam.

The Amateur Extra class license is the highest level of amateur radio licensing. It involves a more rigorous exam covering advanced technical concepts and regulations. Amateur Extra licensees have full access to all amateur bands and modes. This exam has 50 questions. While it's certainly the most challenging, many people find that studying for it deepens their understanding of radio and electronics.

Spectrum and Band Plans

The FCC manages frequency assignments and spectrum band plan considerations for amateur radio in the United States by balancing the needs of amateur operators, other spectrum users, and national interests. The ARRL also represents its members with consultation and

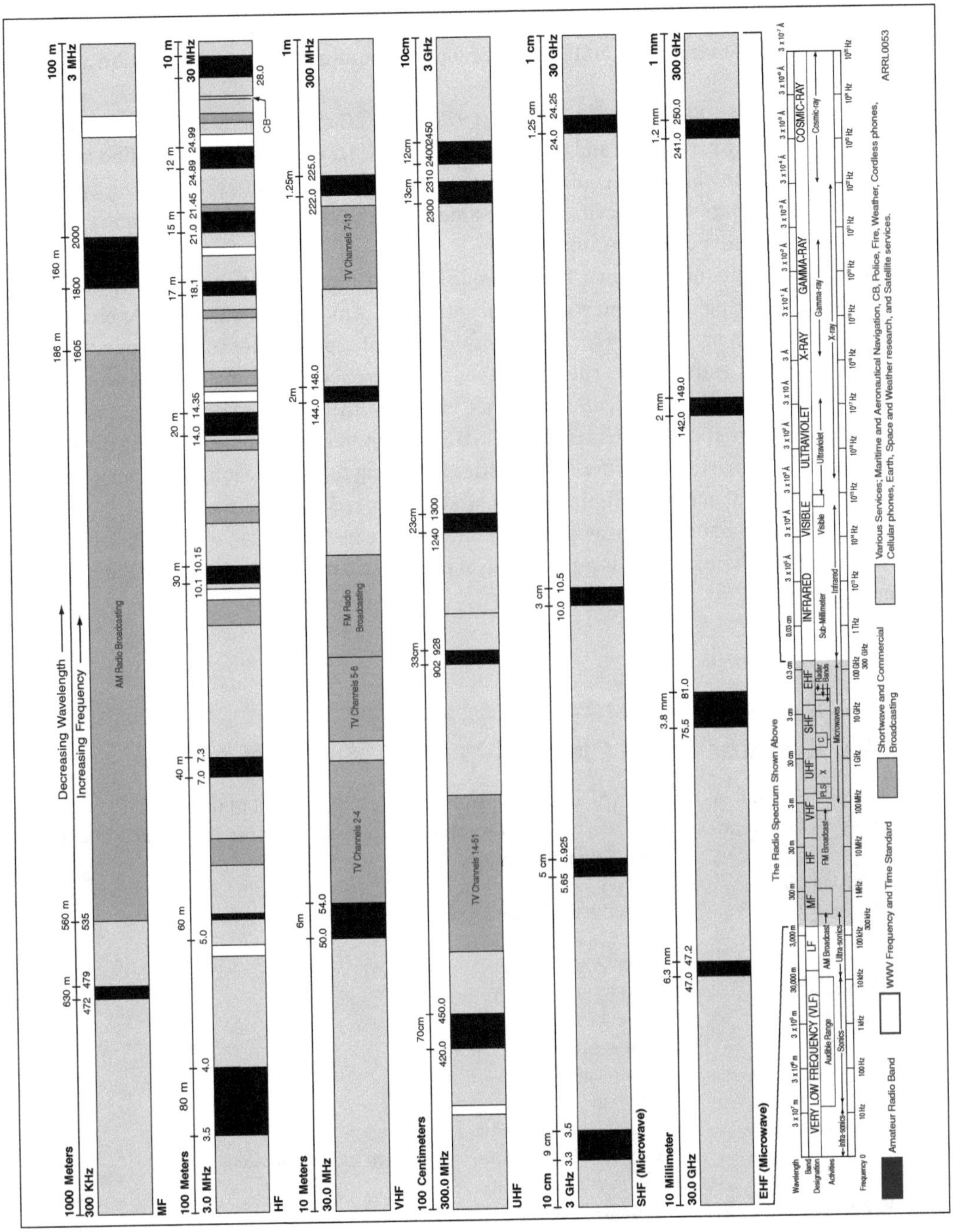

Figure 1.4 — Amateur radio allocations, the black rectangles in the chart, are distributed relatively evenly throughout the radio spectrum.

advisory services. By maintaining flexibility and fostering responsible use, the FCC ensures that amateur radio remains a valuable resource for communication, experimentation, and public service.

Amateur radio operation is governed by Part 97 of the FCC rules, which specifies frequency allocations, operating privileges, and technical standards. Amateur radio is classified as a secondary service in many bands, meaning it must not interfere with primary users (e.g., government, military, or commercial services). Some amateur bands are allocated as primary (e.g., 144 – 148 MHz), granting exclusive amateur use.

The U.S. follows the International Telecommunication Union (ITU) Region 2 band plans for amateur radio, ensuring alignment with global standards. Amateur frequencies are coordinated internationally, particularly for HF bands, to prevent cross-border interference.

The FCC monitors and enforces rules to prevent interference, including action against illegal use of amateur, public service, and broadcast bands by unlicensed operators.

Amateur bands are spread across HF, VHF, UHF, and higher frequencies offering diverse propagation characteristics for different activities including things like long-distance communication, making local contacts, digital modes, and high speed radio networks.

The ARRL 2 meter band plan provides guidelines for the use of the 144 – 148 MHz frequency range in the United States. Below is an example of the ARRL 2 meter band plan for FM mode communications.

ARRL 2 Meter Band Plan (144 – 148 MHz) FM Mode

Please note that this band plan is a general recommendation. Spectrum usage can be different depending on location and regional coordination differences. Please check with your Frequency Coordinator for information.

144.00-144.05	EME (CW)
144.05-144.10	General CW and weak signals
144.10-144.20	EME and weak-signal SSB
144.200	SSB calling frequency
144.20-144.275	General SSB operation
144.275-144.300	Propagation beacons
144.30-144.50	OSCAR subband
144.50-144.60	Linear translator inputs
144.60-144.90	FM repeater inputs
144.90-145.10	Weak signal and FM simplex (145.01, 03, 05, 07, 09 are widely used for packet radio)
145.10-145.20	Linear translator outputs
145.20-145.50	FM repeater outputs
145.50-145.80	Miscellaneous and experimental modes

145.80-146.00	OSCAR subband
146.01-146.37	Repeater inputs
146.40-146.58	Simplex (*)
146.52	National Simplex Calling Frequency
146.61-147.39	Repeater outputs
147.42-147.57	Simplex (*)
147.60-147.99	Repeater inputs

NOTES: (*) Due to differences in regional coordination plans the simplex frequencies listed may be repeater inputs/outputs as well. Please check with local coordinators for further information.

1) Automatic/unattended operations should be conducted on 145.01, 145.03, 145.05, 145.07 and 145.09 MHz.

a) 145.01 should be reserved for inter-LAN use.

b) Use of the remaining frequencies should be determined by local user groups.

2) Additional frequencies within the 2-meter band may be designated for packet radio use by local coordinators.

Notes:

Specific VHF/UHF channels recommended above may not be available in all areas of the US.

Prior to regular packet radio use of any VHF/UHF channel, it is advisable to check with the local frequency coordinator. The decision as to how the available channels are to be used should be based on coordination between local packet radio users.

Some areas use 146.40-146.60 and 147.40-147.60 MHz for either simplex or repeater inputs and outputs.

States use differing channel spacings on the 146-148 MHz band. For further information on which states are currently utilizing which spacing structure, see the Offset Map immediately following.

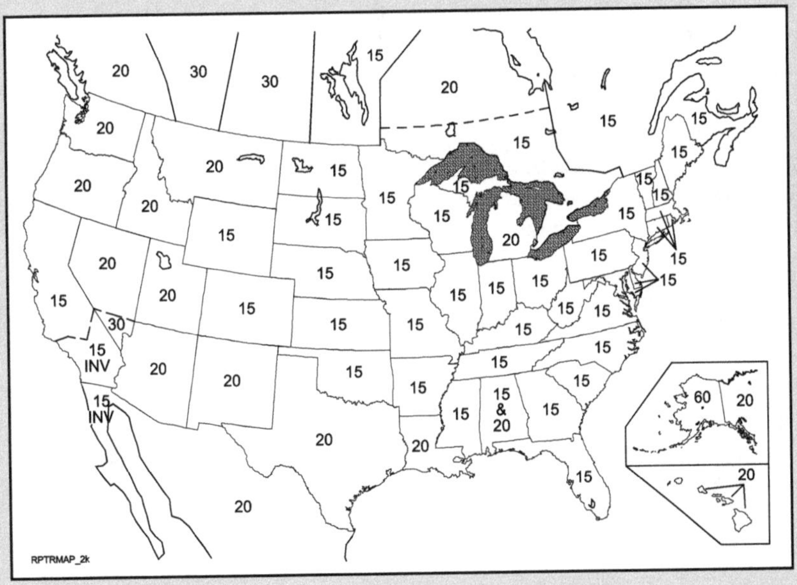

Note: This map shows channel spacing in the US and southern Canada. Spacing is in kHz unless otherwise specified. Please check with your Regional Frequency Coordinator for further information.

ARRL 70 Centimeter Band Plan (420 – 450 MHz) FM Mode

Please note that this band plan is a general recommendation. Spectrum usage can be different depending on location and regional coordination differences. Please check with your Frequency Coordinator for information.

420.00-426.00	ATV repeater or simplex with 421.25 MHz video carrier, control links and experimental
426.00-432.00	ATV simplex with 427.25 MHz video carrier frequency
432.00-432.07	EME
432.07-432.10	Weak signal CW
432.10	Calling frequency
432.10-432.30	Mixed-mode and weak-signal work
432.30-432.40	Propagation beacons
432.40-433.00	Mixed-mode and weak-signal work
433.00-435.00	Auxiliary/repeater links
435.00-438.00	Satellite only (internationally)
438.00-444.00	ATV repeater input with 439.250-MHz video carrier frequency and repeater links
442.00-445.00	Repeater inputs and outputs (local option)
445.00-447.00	Shared by auxiliary and control links, repeaters and simplex (local option)
446.00	National Simplex Callling Frequency
447.00-450.00	Repeater inputs and outputs (local option)

The following packet radio frequency recommendations were adopted by the ARRL Board of Directors in January 1988.

1) 100-kHz bandwidth channels

430.05	430.35	430.65
430.15	430.45	430.85
430.25	430.55	430.95

2) 25-kHz bandwidth channels

431.025	441.000	441.050
440.975	441.025	441.075

Notes:

Specific VHF/UHF channels recommended above may not be available in all areas of the US.

Prior to regular packet radio use of any VHF/UHF channel, it is advisable to check with the local frequency coordinator. The decision as to how the available channels are to be used should be based on coordination between local packet radio users.

Standard offset (difference between receive and transmit) for a 2 meter repeater in the US is plus or minus 600 kHz. The standard for repeaters in the 145 and 146 segments the offset is negative. For repeaters in 147 MHz segment the offset is positive.

UHF repeaters use a plus or minus 5 MHz repeater offset.

1.2 GMRS and FRS

New radio operators often ask: what's the difference between ham radio, GMRS, and FRS? In many cases, GMRS and FRS can easily meet the need for communication. They're ideal for personal, short range use. However, ham radio offers several advantages, such as the ability to communicate across a wide range of frequencies, modes, and specialized services like the Automatic Packet Reporting System (APRS) and satellites. Long-range communication is also possible over HF and with the use of repeaters.

The General Mobile Radio Service (GMRS) is a licensed radio service regulated by the FCC in the United States. It is intended for short-distance, two-way communication and is often used by families, small businesses, and individuals for personal or recreational use without the complexity of amateur radio licensing. GMRS operates in the UHF bands, with specific channels designated for simplex (direct radio-to-radio) and repeater use. Twenty-two channels are shared with the Family Radio Service (FRS), but GMRS has additional repeater-only channels. GMRS requires an FCC license to operate, which costs $35 as of this writing. The license is valid for 10 years and covers the license holder and their immediate family (spouse, children, parents, etc.). No exam is required, unlike amateur radio licensing. Unlike ham radio, GMRS is restricted to voice and tone signaling (no digital modes or data), and license holders are not allowed to modify their radios. In fact, GMRS radios can't even have removable antennas. Radios must be FCC certified, and using a an uncertified radio on GMRS bands is illegal. In fact, it is not legal to use any of the radios discussed in this book on GMRS frequencies. The only legal use for these radios is on the ham radio bands.

The Family Radio Service (FRS) is an unlicensed, two-way, short-distance communication service authorized by the FCC for personal and family use (e.g., hiking, camping, events), groups (e.g., during travel or at amusement parks), and neighborhood and community activities.

FRS is governed by Part 95 of the FCC rules. It operates on twenty-two channels in the UHF band, overlapping with the GMRS frequencies. The radios are limited to power levels of 0.5 to 2 watts effective radiated power (ERP) depending on channel assignment. FRS radios operate in FM mode (frequency modulation) for clear voice communication. Digital features like privacy codes (CTCSS/DCS) are often included to filter out unwanted signals, not to provide actual privacy. Unlike amateur radios, FRS radios are certified by the FCC and cannot be modified. Using an unapproved radio is illegal. As with GMRS, none of the radios discussed in this book are legal to use on FRS frequencies. The only legal use for these radios is on the ham radio bands.

GMRS, FRS, and the Baofeng UV-5R Legal Considerations in the United States

The Baofeng UV-5R is a popular handheld radio, but its use on GMRS or FRS frequencies is not legal. GMRS and FRS radios require FCC certification. To earn that certification the radios must meet certain criteria: they must not be able to transmit out of band and they must not have removable antennas to name just a couple examples. The Baofeng UV-5R does not meet either of these criteria and it is not FCC certified for use on the GMRS or FRS bands.

While the Baofeng UV-5R can operate on both GMRS and FRS frequencies, using it on these bands is illegal. Always check the specific rules for operating, and only use radios that are FCC-certified for these services.

The UV-5R and FCC Certification

The Baofeng UV-5R has had a complicated history with FCC certification. Initially, it was not certified under FCC Part 95, which means it wasn't authorized for use in the General Mobile Radio Service (GMRS) and Family Radio Service (FRS) bands. However, the amateur radio operators can use the radio on amateur radio bands. There are other Baofeng radios that are licensed under FCC Part 95 for use outside of the ham bands, but that is somewhat outside the scope of this book, as our focus is primarily on ham radio.

1.3 Baofengs Outside the Ham Bands

A warning to Baofeng users regarding out-of-ham-band transmissions: some versions of the UV-5R and other models in the product line are capable of transmitting and receiving outside the ham bands. While it's legal to listen to frequencies outside the ham bands, the FCC has not approved these radios for operation in business or land mobile bands. Do not program these frequencies for transmission. Listening to these channels can be useful for monitoring local public safety channels, but operating illegally could result in an FCC citation, or a visit from an FCC inspector. I will cover how to properly set up these radios' receive-only channels to help avoid accidental transmissions in chapters 6 and 7.

Baofeng radios can also produce illegal harmonics that violate FCC rules. These unwanted emissions can cause interference with other services, potentially leading to complaints and FCC enforcement actions. Because these are ham radios rather than GMRS or FRS radios, the manufacturer is not responsible for the out of band emissions, the operator is. That means it's a good idea check the harmonics being emitted by your radio to make sure they are in line with FCC regulations. You can also seek out FCC compliant versions of these radios, though unscrupulous distributors have been known to mislabel the radios they're selling. It is always best to purchase equipment from a reputable ham radio distributor.

1.4 A Word about Handheld Radios

Handheld radios are great for portability, affordability, and getting started, while mobile radios and even larger base station radios offer increased range due to higher power outputs and the use of external, higher gain antennas. While mobile radios do require installation of the radio in a vehicle or at a stationary operating position, which can be more complicated, many users appreciate the increased power and reliability of these more permanent installations.

That said, there's nothing quite like programming a repeater frequency into your UV-5R right out of the box and making contacts.

Table 1.1

Handheld versus Mobile Transceiver Comparison

	Handheld Transceivers	*Mobile Transceivers*
Portability	Highly portable, lightweight, easy to carry	Fixed installation limits portability
Power Output	Low (typically 2 - 10 watts)	Moderate (25 - 65 watts)
Range	Limited due to lower power and shorter antennas	Extended range due to higher power and better antennas
Battery Life	Limited by internal battery packs and their capacity	Operates on vehicle or external power with no limitations
Antenna Options	Limited; depends on stock or upgraded whips	Flexible; uses external or vehicle-mounted antennas
Performance	Limited performance	Better performance providing band pass filtering, enhanced displays, and features
Cost	$–$$	$$–$$$

Antenna Gain

The gain of an antenna is related to its directivity, and directivity is related to the shape of the directive (or radiation) pattern, which I discussed in the ERP sidebar. Because directivity is based solely on the shape of the directive pattern, it does not consider any power losses that may occur in an antenna system. To determine gain, these losses must be subtracted from the power supplied to the antenna. In many antenna systems that amateur radio operators use, the system's efficiency is quite high, making the gain essentially equal to the directivity. However, this isn't the case for rubber duck antennas. These antennas incur relatively high losses. More efficient antennas like quarter or half-wavelength dipoles and Yagis (see the Antennas section of Chapter 9) have greater directivity and lower losses, increasing their gain considerably when compared to rubber duck antennas.

2 Simplex Transmissions

2.1 Practical Applications and Use Cases

Hams enjoy talking on their radios for many reasons, and the topics they discuss can cover just about everything. Some like to converse during their daily commute or while running errands, while others like participating in special interest radio nets. Many also make it a point to keep in contact with club and family members. Whether from the comfort of home, in vehicles operating mobile, outdoors on a farm, or hiking up a mountainside, ham radio can be found, used, and heard everywhere.

Using simplex (direct radio-to-radio) on handheld VHF/UHF radios is great for communicating within a local area. It's as fast as pushing your push-to-talk (PTT) button, saying what you need to say, and then listening for the response. Occasionally, you might contact someone on a repeater and discover they're within simplex range. When possible, the polite thing to do in this situation is clear the repeater so that others can use it. According to FCC rules, hams are supposed to put out the minimum power necessary to make their contact. Using a repeater to blast out a message you could just as easily send with your handheld may not violate the letter of the regulation, but it sure seems out of touch with its spirit.

It's also important to have realistic expectations for how far you can communicate with a handheld transceiver. Your antenna, power level, height above average terrain, and other factors all impact how far your signal can travel. Without upgrading certain aspects of your Baofeng, you'll likely be limited to transmitting no farther than line-of-sight, which is typically between 5 and 15 miles. If you are operating in a forest or a densely populated city, the effects of foliage or buildings on VHF and UHF signals can be quite detrimental, limiting your range of communication even more. So, with a handheld radio, don't expect to hike for days into the wilderness and be able to communicate with civilization — or even a base camp on the other side of a steep ridge. That said, there are tons of things that you can do with a handheld radio. The following sections present some of the most common ways hams are having fun with their handhelds on simplex frequencies.

2.1.1 Training

Using handheld radios in simplex mode is a popular choice for many ham radio operators and clubs when teaching and training newcomers to the hobby. Simplex communication is an excellent first step for learning how to make contacts, manage frequencies, and understand basic propagation.

This approach helps alleviate the "first contact" anxiety that many hams experience when using a repeater with a wide coverage area and large audience. Additionally, organizations like (Amateur Radio EmergencyService (ARES) and RACES (Radio Amateur Civil Emergency Service (RACES) regularly conduct drills to keep operators prepared and proficient in emergency communication skills.

2.1.2 Local Radio Nets

Some amateur radio clubs host simplex nets for casual on-air gatherings, allowing local operators to communicate directly with each other on VHF and UHF frequencies. These nets foster a sense of community and friendship among local hams. Special interest topics often drive engaging conversations during these nets, making them both educational and enjoyable.

2.1.3 Supporting First Responders

Many hams are aware of the challenges that arise during power outages, natural disasters, or other emergencies where simplex communication can become a critical backup when repeaters are unavailable. Handheld radios can achieve communication over distances of up to 1 to 2 miles or more (5 to 25 miles) under optimal conditions, particularly when paired with an external amplifier and a high-gain antenna. This capability was demonstrated in after-action reports from operator

Amateur Radio Net (Ham Net)

An amateur radio net, or more simply a ham net, is an "on-the-air" gathering of amateur radio operators. Most nets convene on a regular schedule and specific frequency, and are organized for particular purposes by radio club members including relaying messages, discussing a common topic of interest, severe weather (for example, during a SKYWARN activation), emergencies, or simply as a regular gathering of friends for conversation.

Figure 2.1 — Assessing damage in Hahnville, LA, during the aftermath of Hurricane Ida. [Photo courtesy of FEMA]

Thomas Witherspoon, K4SWL in the Asheville, NC area after Hurricane Helene in 2024, where ham radio played a vital role in maintaining community contact and supporting restoration and medical efforts.

2.1.4 Public Service

The ARES, RACES, and Radio Emergency Associated Communication Teams (REACT) often rely on simplex communications to support local emergency responders. In emergencies where cellular and internet networks are unavailable, amateur radio volunteers assist first responders and the Federal Emergency Management Agency (FEMA) by providing critical communication lifelines. Many amateur radio clubs also have public service and community outreach programs, where volunteers offer communication support and manpower at events like parades, races, and other community projects. These teams frequently use handheld radios, which provide reliable communication over short distances, typically 1 to 2 miles depending on terrain and antenna height. With experience, operators can improve coverage (5 to 25 miles) and reliability by using mobile antennas or mounting antennas 20-feet or higher in the air. These public service events also serve as valuable opportunities for operators to practice and refine their communication skills, preparing them for real emergency situations. Sometimes the teams use the CTCSS tone (usually 100 Hz) for tactical purposes or to reduce interference.

Public safety teams, along with ham radio clubs, hold emergency tests throughout the year to practice their skills and be better prepared when it really counts.

2.1.5 Fox Hunts

I've organized several "fox hunts" (hidden transmitter hunts) for club members and Boy Scout groups, where participants use handheld radios in simplex mode with Yagi antennas that they've built themselves from chopped up tape measures to locate hidden transmitters. These activities are an excellent way to improve radio skills while having fun. To add an extra layer of excitement, fox hunts can be combined with scavenger hunts, using multiple hidden transmitters in a park or along a trail. Participants not only locate the transmitters but also gather prizes in a "geocaching-style" activity. These events provide a practical introduction to direction-finding techniques, which are valuable skills for locating transmitters in various scenarios. This is a very popular club event.

Figure 2.2 — Hams and non-hams of all ages can enjoy foxhunting.

2.1.6 POTA/SOTA

Figure 2.3—These hams summitted Cinnamon Butte for a ladies-only SOTA activation. Mount Thielsen can be seen in the background.

Portable radio operators and handheld radios are essential for Summits on the Air (SOTA) and Parks on the Air (POTA) activations. Operators often use handheld radios with the standard antenna, but a longer 16" whip or lightweight portable antenna provides better range to make contacts from remote locations such as mountain summits or parks. In group efforts with other operators at multiple parks, operators often compete for bragging rights by trying to make the most distant contact. The Colorado 14er event is a yearly event where operators in the Rocky Mountain regions provide abundant opportunities for summit-to-summit contacts, which are especially rewarding.

Two particular SOTA activations stand out as memorable for me. From Mount Diablo (CA), I made a 150-mile contact with a station in Nevada on 2-meter FM, with 5 watts, and a high gain whip antenna. Another summit in Maine allowed me to contact several mobile stations driving along a nearby highway, including an unexpected and delightful contact with a station that I had previously contacted several times from my home park on HF.

2.1.7 The Outdoors

Hikers, campers, and outdoor enthusiasts often rely on handheld VHF/UHF radios to stay connected in remote areas where cellular coverage is unreliable. The ability to use higher power and external antennas significantly increases range and improves communication reliability — capabilities not possible with FRS radios, making ham radio a superior choice for group communications. The ability to carry additional battery packs in a backpack can extend operating time. Many hams use the North America Adventure Frequency (NAAF) of 146.58 MHz as a designated channel for outdoor activities, as well as a meeting point for POTA and SOTA activators, ensuring the 2-meter national calling frequency (146.52 MHz) remains clear for its intended use.

Figure 2.4 — I'm kitted out for some winter woods time.

On fly-fishing and ice-fishing trips, I have used the NAAF to stay in contact and share updates about our activities with my fellow ham fishermen. For extended trips, I often set up external antennas to enhance our range and ensure our signals can travel across lakes or cover larger areas effectively.

2.1.8 ARRL Field Day/Contests

During ARRL Field Day operations and other contests, operators often use handheld radios in simplex mode to make contacts. Belt clipped radios provide internal communications and status reports. The ARRL sponsors several VHF and UHF contests, where competitive amateurs aim to make as many contacts as possible. These events are not only fun but also serve as great practice for improving operating techniques in various conditions. Using a Yagi antenna gives you greater directionality and gain, and can increase the distance over which you can communicate.

2.1.9 Boy Scouts of America

Figure 2.5—Boy Scouts have honed their radio skills for decades. This photo dates to December 15, 1923, and features Hiram Percy Maxim, 1AW, at the radio.

The Boy Scouts of America offer a radio badge that serves as a gateway to the world of ham radio, introducing young enthusiasts to the basics of communication and radio operations. This badge encourages Scouts to learn about radio technology, operate radios, and understand the importance of communication in emergency and community service situations. Baofeng radios, with their affordability and accessibility, make an excellent choice for Scouts working toward this badge. These handheld transceivers provide a cost-effective entry point into ham radio, allowing Scouts to gain hands-on experience with VHF/UHF frequencies and simplex communication, essential skills for both the badge and further ham radio pursuits.

2.1.10 Automatic Packet Reporting System

The automatic packet reporting system (APRS) is a digital communication system used by amateur radio operators to exchange messages and track locations using GPS. Widely recognized in the ham community, APRS has numerous applications in emergency communications, tracking, and weather reporting. The standard operating frequency for APRS is 144.390 MHz. Radios equipped with APRS transmit location and text information over the airwaves, which other APRS stations receive and display on websites like **aprs.fi**. High-altitude balloon

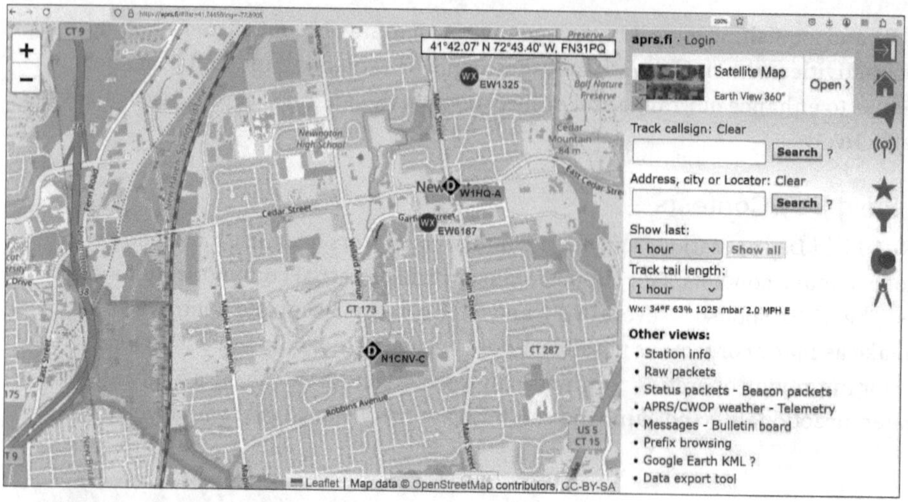

Figure 2.6 — An example of a map populated by APRS from aprs.fi centered on Newington, CT.

experiments often incorporate APRS payloads to track the balloons during their journeys. Some even navigate around the globe. APRS can be operated using an external APRS terminal node controller (TNC) modem.

2.1.11 Off Roaders and Over Landers

When outside of local repeater coverage areas, simplex channels are the only way to communicate. By using the included rubber duck antenna or connecting to a mobile antenna, vehicle-to-vehicle communication becomes more reliable. Off roaders and over landers have found that ham radio offers a more dependable communication method than GMRS or FRS, especially for coordinating travel routes, alerting others about hazards, or sharing updates over short to medium distances. Ham radio's versatility and reliability make it a valuable tool for these outdoor adventures.

Jeep 4×4 enthusiasts, along with other off-roaders, often use 146.460 MHz as a simplex frequency for vehicle-to-vehicle communication. This is an informal, nationwide simplex channel. It allows for reliable communication over short to medium distances when outside of repeater coverage areas. As with any frequency, it's important to follow the guidelines and etiquette for amateur radio use.

Don't forget the NAAF frequency, which is also used for outdoor adventure communications.

2.1.12 SKYWARN®

The National Weather Service (NWS) utilizes amateur radio as one of its primary methods for communicating with spotter groups and emergency management organizations within their areas of responsibility. For decades, amateur radio operators have played a vital role in supporting the SKYWARN storm spotter program, using their unique communication capabilities to relay critical information between the NWS, local emergency management officials, and storm spotter networks. Weather forecasting and radio communication classes help ham radio operators enhance their knowledge and engage in community service. Simplex communication is particularly important, as local repeaters may sometimes be out of service during severe weather events. It's a good practice to program all 7 NWS channels (162.400 MHz, 162.425 MHz, 162.450 MHz, 162.475 MHz, 162.500 MHz, 162.525 MHz, and 162.550 MHz) into your radio to receive weather reports no matter where you are. I'll cover how to do this in chapter 6.

2.1.13 Travel Radio

Amateur radio operators, like myself, often use handheld radios in their vehicles for on-the-go communication, and simplex operation is an excellent option for short-distance communication while traveling. I find this particularly useful on road trips where repeaters aren't always available. For example, I can talk to others on 146.52 MHz (typically called Five Two) simplex, as many mobile operators monitor the national calling channels. Using a speaker microphone, a 12 V dc battery eliminator, and a magnetic mount antenna enhances the overall experience.

The UV-5R is small, light, and compact, and with an extra battery or two, it fits easily in my carry-on bag for quick deployment in a rental car. I can't count how many times I've enjoyed being mobile with my handheld and having conversations with local hams while on business travel, from the airport in Boston to Salem, NH, and then off to Hartford, CT. Not only was I able to make new radio friends, but I also got real-time travel information and discovered local events. Having conversations on repeaters along the highway is also rewarding. For example, I was talking to another mobile operator and as we got out of line-of-sight range he suggested we move to the W1BRS repeater located in Manchester, CT. He suggested a must-see was the USS Nautilus submarine in Groton, CT.

The *ARRL Repeater Directory* was also invaluable to me for finding repeater frequencies along my route. I set up a *CHIRP* travel file with all the frequencies for a quick upload, and for a while, I kept a second Baofeng labeled with "MA, NH, and CT" packed in a travel go-kit that was stored in my suitcase for convenience. The Baofeng, with its 128 memory locations, makes this easy.

2.1.14 Antenna Experimentation

For those who enjoy building antennas and testing different setups, handheld radios are perfect for experimenting with and testing new antenna designs in real-world environments. Simplex VHF/UHF communications are especially useful for examining signal paths, band conditions, and how factors like terrain, weather, or atmospheric conditions affect transmissions using different antennas.

2.1.15 Simplex Wrap Up

In summary, handheld radios in simplex mode offer many versatile options for local communication, emergency preparedness, portable operations, and social activities. However, it's crucial to use these radios responsibly and within the legal operating parameters to avoid potential issues with the FCC.

3 Repeaters

3.1 How Repeaters Work

A ham radio repeater is a system that extends the range of two-way radio communication by receiving a signal on one frequency and retransmitting it on another, often with increased power and from an elevated antenna location, such as a tall tower, hill, or building. Repeaters are typically designed with coverage footprints of 30 to 50 miles, but high point repeaters can offer a much greater coverage area. Repeaters are commonly used on the VHF and UHF amateur radio bands, and all of the radios covered in this book allow you to access them. Frequency pairs for repeaters are typically assigned and coordinated by local volunteer groups called Frequency Coordinators, who can be a great resource when setting up your radio. Many repeaters also require a Continuous Tone-Coded Squelch System (CTCSS) tone, often simply referred to as a "tone," to access the system. I will discuss tones in more detail in the next section.

Repeaters listen for signals from user radios on the input or uplink frequency then retransmit the received signal on the output or downlink frequency. When programming the repeater (or looking at a repeater listing) the output or downlink frequency is the one on which you listen. You need to program a transmit offset frequency, which is the difference between these frequencies (typically +/- 600 kHz for VHF, or +/- 5 MHz for UHF). The offset is required to allow a repeater to receive a signal and retransmit it simultaneously. If a repeater didn't use an offset, it would be receiving and transmitting on the same frequency, causing interference and making both signals unintelligible. Repeaters use a higher power than most handheld transceivers and gain antennas to achieve better coverage. These systems are also designed for continuous use and feature robust components and often battery back up systems so they can continue operating during power outages. Some clubs operate solar powered repeaters in difficult to reach locations, increasing the repeater's ability to maintain broad, reliable communications capabilities.

Many repeaters provide a courtesy tone or beep that is transmitted after the repeated communication. This beep also signals that when you key up your radio, the repeater responds with a carrier. When you stop transmitting, the repeater pauses briefly to determine if your

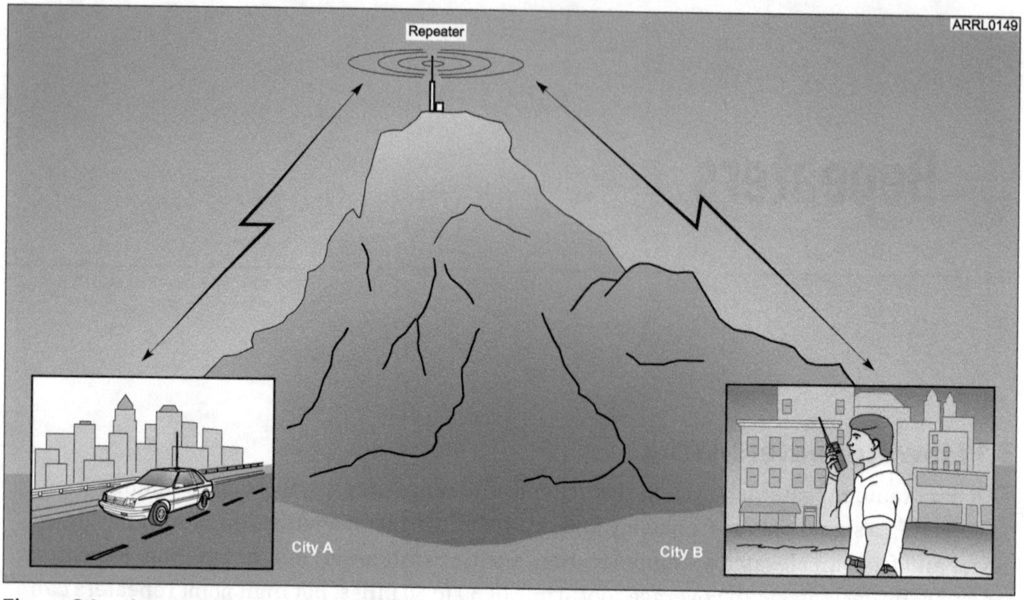

Figure 3.1 — A repeater extends the range of its users, allowing them to communicate over longer distances.

transmission was momentarily interrupted. If it was, the repeater allows you to continue transmitting once it detects your continuing signal. Otherwise, it sounds the courtesy beep, indicating the end of the repeater's retransmission. The repeater then switches from transmit back to receive and waits for the next transmission.

If you don't wait to hear the courtesy tone before keying up your transceiver, you run the risk of overrunning the repeater's time-out timer. Many repeaters have time-out timers which automatically terminate your transmission if it runs too long. The timers usually give you 60 or 90 seconds to speak. For example, the W1BOS repeater in Boston is set at 30 seconds, so transmissions on that repeater need to be very short. When you do overrun the time-out timer, the repeater shuts down, and everyone trying to access it must wait for it to reset. The repeater then plays a message, such as, "repeater time out." When in the middle of a net or conversation, this can be annoying. Don't be afraid to ask the local repeater users what the length of the time-out setting is. Waiting to hear the courtesy beep before transmitting allows the timer to reset, making repeater operation smoother. Courtesy beeps are also used by blind amateurs to identify which repeater they are using.

Repeaters do not allow multiple simultaneous inputs, which means only one person can talk at a time. When two people try to talk at the same time, the person with the stronger signal captures the repeater and the other user remains unheard, or the two signals interfere with each other. This is called doubling, and can be avoided by listening before transmitting, or if it's a net, the net control station will call out to users when it's their turn to talk. This is called rotation and is always used when three or more people are talking to ensure everyone has a place in line and a chance to speak.

Recently, new portable repeaters have appeared on the market at affordable prices. These are deployed for tactical purposes at public service events, or for emergency communications support. Some clubs have built communication trailers for their club activities; having a portable or temporary repeater allows the club to extend its communication range. If you are a member of these groups, you may wantto make sure these portable repeaters are programmed in your radio so that when they are on the air you are ready to communicate.

3.2 Continuous Tone-Coded Squelch System

Continuous Tone-Coded Squelch System (CTCSS), also known as "tone squelch," is used on repeaters to reduce unwanted transmissions and ensure the repeater only responds to signals carrying the correct sub-audible tone. This tone, a low-frequency signal added to the radio transmission, acts as a "key" to activate the repeater. Only radios transmitting the correct CTCSS tone can access and trigger the repeater.

CTCSS tones can also be applied to the repeater's output signal to help users filter out interference in crowded environments. In this case, the repeater transmits a specific CTCSS tone along with its output signal. If your receiver is programmed to only open its squelch when it detects this tone, you will only hear transmissions from the repeater and can block out other signals or interference on the same frequency.

This feature is particularly useful in areas with overlapping repeaters on the same frequency or in regions with high levels of VHF/UHF activity. By using CTCSS, users can maintain clearer communication and reduce the distraction of unwanted noise or transmissions.

Digital coded squelch (DCS) is a method of access control and noise filtering used in ham radio and other communication systems. Like CTCSS, DCS allows a repeater or radio to ignore unwanted signals by requiring a specific digital code for activation. Instead of using a sub-audible tone like CTCSS, DCS transmits a digital code along with the signal. DCS usage is much less common than CTCSS, but both protocols provide the same result.

3.2.1 How CTCSS Works with Repeaters

To access a repeater, users must program their radios to transmit the proper CTCSS tone. Additionally, enabling "tone squelch" on the radio ensures that users only hear transmissions from the repeater and not background noise or stray interference.

New operators often encounter issues accessing a repeater because they haven't programmed their radio with the correct CTCSS tone. To avoid this, consult resources like *The ARRL Repeater Directory* or online repeater databases like **repeaterbook.com** to find the repeater's input and output frequencies, as well as the required CTCSS tone, before programming your radio.

Many new hams have unintentionally missed nets or other activities because they overlooked this step, so always confirm the required settings in advance to ensure successful communication.

Let's look at a practical example for the W1AW repeater located in Newington, CT.

Output Frequency: 145.450 – This is the frequency programmed into our radio to receive the repeater's signal.

Shift: -0.600 MHz – The repeater receives incoming transmissions that are 600 kHz lower than the output frequency. The radio must transmit on this frequency.

Input Tone: 127.3 Hz CTCSS

Output Tone: 127.3 Hz CTCSS

When you press the radio's push-to-talk (PTT) button, the repeater picks up the signal, verifies the 127.3 CTCSS tone, and then retransmits the signal on the output frequency. If the repeater uses an output tone, you must also program it into the memory location.

Table 3.1
CTCSS: Codes

The standard Electronic Industries Association (EIA) tones, in hertz, with their Motorola alphanumeric designators, are as follows:

67.0 – XZ	97.4 – ZB	141.3 - 4A	210.7 - M2
69.3 – WZ	100.0 - 1Z	146.2 - 4B	218.1 - M3
71.9 – XA	103.5 - 1A	151.4 - 5Z	225.7 - M4
74.4 – WA	107.2 - 1B	156.7 - 5A	233.6 - M5
77.0 – XB	110.9 - 2Z	162.2 - 5B	241.8 - M6
79.7 – WB	114.8 - 2A	167.9 - 6Z	250.3 - M7
82.5 – YZ	118.8 - 2B	173.8 - 6A	
85.4 – YA	123.0 - 3Z	179.9 - 6B	
88.5 – YB	127.3 - 3A	186.2 - 7Z	
91.5 – ZZ	131.8 - 3B	192.8 - 7A	
94.8 – ZA	136.5 - 4Z	203.5 - M1	

Some systems use tones not listed in the EIA standard. Motorola designators have been assigned to the most commonly used of these tones: 206.5 (8Z), 229.1 (9Z), and 254.1 (0Z). Some newer amateur transceivers support additional tones of 159.8, 165.5, 171.3, 177.3, 183.5, 189.9, 196.6, and 199.5 hertz.

3.3 Linked Repeaters

Most repeaters are standalone devices, providing their individual pool of coverage and nothing else. But a significant number of repeaters are linked — connected to one or more other repeaters. Those other repeaters can be on other bands at the same location, or they can be in other locations, or both. Linked repeaters let users communicate between different bands and across wider geographic areas than they can on a single repeater.

There are many ways to link repeaters. Repeaters on the same tower can just be wired together, or they may even share the same controller. Repeaters within a hundred miles or so of each other can use a radio link – separate link transmitters and receivers at each repeater with antennas pointed at each other. Multiple repeaters, each still about 100 miles apart, can daisy chain their links to cover even wider territory. There are a few linked repeater systems in the United States that cover several states with dozens of repeaters, but most radio-linked repeater

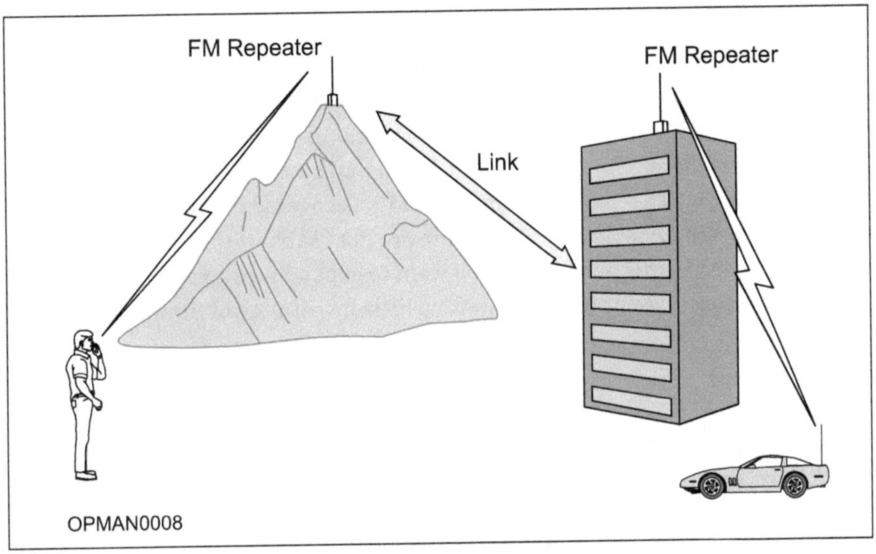

Figure 3.2 — Repeater linking can greatly expand VHF/UHF communication distances. Repeater links are commonly made via dedicated radio hardware or via the internet.

systems have more modest ambitions, covering just part of one or two states. The PAPA System is an example of a linked repeater system. It covers an area from Fresno to San Diego in California and out to Las Vegas in Nevada.

Repeater linking via the internet makes it possible to tie repeaters together around the world and in nearly unlimited number. I'll talk more about internet linking in the next section, and I'll go into how to use a Baofeng DMR handheld, the DR-1801UV, to connect to the internet-linked DMR network in chapter 8.

There are several ways linked repeaters can be operated, coming under the categories of *full time* and *on demand.* Full time linked repeaters operate just as the name implies — all repeaters in a linked network are connected all the time. If you key up one of them, you're heard on all of them, and you can talk to anyone on any of the other repeaters on the network at any time. You don't have to do anything special to activate the network, since it's always there.

In an on-demand system, the linked repeaters remain isolated unless you take some action, usually by sending a code by DTMF digits (see chapter 7, specifically section 7.4.3, Step 7, DTMF for more information), to connect them. Your DTMF sequence may activate all the repeaters on the network, or the system may let you address just one specific repeater, somewhat like dialing a telephone. When you're finished, another DTMF code drops the link, or a timer may handle that chore when the repeaters are no longer in use.

3.4 Internet Linking

The internet and *Voice over Internet Protocol* (VoIP) has expanded repeater linking exponentially, making worldwide communication through a local repeater commonplace. Two internet linking systems, *IRLP* and *EchoLink®*, have reached critical mass in the United States and are available almost everywhere. The D-STAR and DMR digital systems also have a significant internet linking component. This topic, specifically DMR, is covered in greater detail in Chapter 8. The internet connected systems mentioned here are not usable with a Baofeng DR-1801UV, the digital radio covered in this book, but if you want to expand your experience in digital radio, these are interesting systems to explore with a little investment of time and money.

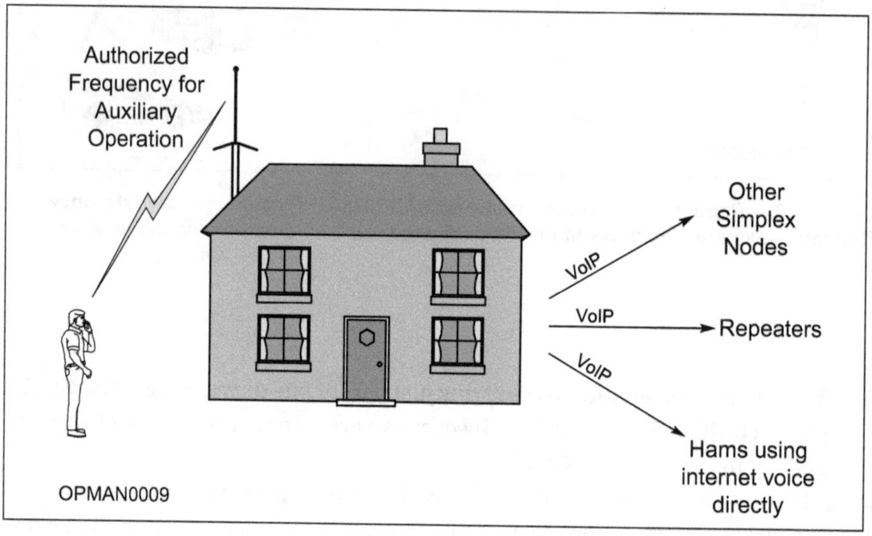

Figure 3.3 — A diagram of a VoIP simplex node.

3.4.1 IRLP

The *Internet Radio Linking Project (IRLP)* is the most "radio" based linking system. User access is only via radio, using either simplex or repeater stations, while linking is done using VoIP on the internet. *IRLP* system operators establish a node by interfacing their radio equipment to a computer with an internet connection and then running *IRLP* software. Once that's set up, repeater users send DTMF tones to make connections either directly to other individual repeaters or simplex nodes or to *reflectors* — servers that tie multiple nodes together as one big party line.

The direct connections work like on demand linked repeaters. You dial in the node number you want to connect to and access code (if required), and you are connected to the distant repeater. Once connected, everyone on both ends can communicate. When finished, another

DTMF sequence takes the link down. Someone from a distant repeater can make a connection to you as well.

Reflectors work like a hybrid between on-demand and full-time linked repeaters. You can connect your local repeater to a reflector and leave it there all day, or you can connect for a special purpose (like a net) and drop it when the event is over.

3.4.2 EchoLink

EchoLink requires a PC with a sound card and appropriate software. It allows repeater connections like IRLP, and it has Conference Servers, similar to IRLP reflectors that permit multiple connections. The big difference is that *EchoLink* allows individuals to connect to the network from their computers without using a radio.

The *EchoLink* conference servers all have more or less specific functions. Some are just regional gathering places, while some are region, topic, or activity based (SKYWARN® and National Hurricane Center Nets, Jamboree on the Air, and so on).

You can connect your *EchoLink*-enabled computer to a digital capable base station radio fairly easily through a sound card and create an on-air node. Don't pipe *EchoLink* to a local repeater without permission from a repeater owner though.

If you decide to create a full-time link from a computer to a repeater, consider using a dedicated UHF link frequency rather than just a base station on the repeater input. This applies to IRLP connections as well. Of course, the internet is infrastructure dependent, and both power and internet access can be interrupted during storms or other disasters. You can also acess EchoLink from a radio repeater using DTMF commands. For example, the # key disconnects you from the network.

3.5 Repeater Etiquette

If there is one area of concern after a new ham nervously fumbles through his first few contacts on the repeater, it's etiquette.

Repeater operation is straightforward, and you learn to confidently make contacts as you listen to others. That said, many repeater groups have rules you must follow. Always check with the repeater owners regarding any rules they have for their repeater operation.

Remember, club repeaters are usually monitored by ham radio users and scanner listeners of all ages. Common sense and professional behavior is expected. Hams are encouraged to operate respectfully and legally. Here are some "best behaviors" to follow when using a repeater.

- Always listen before transmitting. This is to avoid interrupting ongoing conversations, unless it's a true emergency. Don't just turn the radio on and call out without first listening to make sure the repeater is idle.
- It is also important to leave a 2 to 3 second pause between transmissions during a conversation. (Often this behavior is enforced by the courtesy beep, though not all repeaters employ them.) This allows others to join, including stations with emergencies or those needing to use the repeater briefly. I have been guilty of using a

fast trigger, which prevents others who would like to join the conversation from doing so.

- Avoid using the term "BREAK" or "BREAK BREAK," as this is the standard term used to break into a conversation to pass emergency traffic. Another to avoid is the term "emergency traffic," which is used to let the user know that someone needs to communicate about an emergency. The proper way to join an ongoing conversation when there is no emergency is to state your call sign during the gap period and wait for the other users to acknowledge and welcome you to the conversation.
- Hams must also identify themselves properly with their assigned call sign. Begin or end your communication with your call sign. For example, "This is VE3IPS monitoring," or "W1AW, W1AW, this is John, VE3IPS calling. Are you listening?" The FCC requires that you identify yourself at least once every 10 minutes during a conversation and at the end of your final transmission.
- Keep transmissions short to allow others to join in. Long-winded transmissions can monopolize the repeater and may be frowned upon. As discussed above, this also prevents repeater hogging, especially during the busiest peak "commute-hours" usage of the repeater. Most users switch to a simplex channel to continue their conversation, allowing the repeater to be available for other users.
- Speak clearly into the microphone, holding it about an inch away from your mouth. This ensures a proper signal level. You may find others whose audio is muffled or distorted from talking too loud or too close. You may hear the phrase "eating the microphone" for those that have bad audio. Radio checks are used to determine optimum settings for your voice. Set this up using a simplex channel.
- Be respectful and courteous. It is best to avoid controversial or sensitive subjects like politics and religion unless you're certain it's appropriate for the audience. These are public airwaves, so it's essential to avoid arguing or using inappropriate language.
- Announce your operation condition by stating "VE3IPS Clear" which means "I have finished my communication with W1AW and I am listening." You can also say "VE3IPS monitoring" which informs other listeners that you are listening and available for conversations. You can also say you are "QRT," which lets users know you are no longer using the repeater.
- It is illegal and frustrating to conduct a radio check without identifying. Transmissions without identification often occur when an operator wants to see if they can access the repeater from their location. Transmitting on a repeater without identifying is called "kerchunking." The repeater responds with a pause and a courtesy tone, proving that you can access the repeater. Some users kerchunk several times, but it's proper etiquette to identify yourself and state your purpose. When testing a new radio on a repeater, I say some version of, "VE3IPS testing for access," or even "VE3IPS range check," which is better than dead audio. Better still is, "This is VE3IPS radio check please. Is anyone listening that can provide a signal and audio report?"

The Amateur's Code

The ARRL (American Radio Relay League) has a set of guidelines known as the Amateur's Code that outlines the ethical standards for ham radio operators. It is:

Considerate...never knowingly operate in a way that diminishes the pleasure of others.

Loyal...offer loyalty, encouragement, and support to other amateurs, local clubs, and the IARU Radio Society in your country.

Progressive...keep your station up-to-date, well-built, and efficient. Operate with practices above reproach.

Friendly...operate slowly and patiently when requested; offer friendly advice and counsel to beginners.

Balanced...ensure radio is a hobby that never interferes with duties owed to family, job, school, or community.

Patriotic...keep your station and skills ready for service to your country and community.

The original Amateur's Code was written by Paul M. Segal, W9EEA, in 1928.

- Don't call CQ on the repeater. A simple, "VE3IPS monitoring" tells listeners that you are listening and ready to have conversations. When you're listening on a repeater, you may make this announcement every ten minutes or so, allowing interested hams to call you.

By following these guidelines, you contribute to an enjoyable and efficient operating environment for everyone using the repeater.

3.6 Repeater Resources

Before you can operate using repeaters to extend your range, you need to know the repeater details. Radio clubs often post their repeater details on the club website. There are also various online resources. In some areas there are statewide repeater groups like the Cactus Intertie System, PAPA Systems, or the CT ARES group that has comprehensive interlinked repeater resources.

The most popular online resource is **repeaterook.com**, though others do exist. You should choose whichever suits your needs best. **Repeaterbook.com** offers a smartphone app that is intuitive and makes discovering repeaters easy, especially when travelling to a new location or looking for new repeaters that are being activated in your area. The database and app provide a wealth of information that can assist in programming repeaters into memory on your radio. The site provides information on coverage and special repeater features, like the repeater's primary mode and whether it's digital-mode capable. Some repeaters even provide hotlines and weather alert broadcasts.

Figure 3.4 — The RepeaterBook website.

Active hams often have a radio set to their busy local repeater to listen in on traffic reports and daily chit-chat on a wide variety of subjects. It's easy to have several UV-5Rs around the house to listen in and join in conversations.

If you prefer analog content, like books, to smartphone apps and online databases, you might check out the *ARRL Repeater Directory* book. There are more than 22,300 repeaters in the United States and Canada. The best way to find repeaters in your area, when you're traveling, or when you're on the move, is with the *ARRL Repeater Directory.* Look up repeaters and see their input and output frequencies, CTCSS, and repeater shift, plus find out whether they are analog FM repeaters, or one of the varieties of digital machines. Band plans are also included in a convenient 6 × 9" spiral bound format. Having a paper copy of a repeater guide is ideal when your electronic resources or internet networks aren't available.

Figure 3.5 — The ARRL Repeater Directory®.

The *ARRL Repeater Directory* includes crowdsourced listings contributed by users, repeater owners, and volunteer frequency coordinators. Portable operators and public service volunteers often keep a copy nearby or in their go-kit.

These resources are ideal for road trip planning and having several memory files that can be loaded into the radio as needed. The Baofeng has a lot of memory positions, so adding repeaters for these uses is easy.

4 Bug-Out Bags and Go-Kits

4.1 The Difference between Bug-Out Bags and Go-Kits

Though bug-out bags and go-kits serve different purposes, they can overlap depending on the situation. A bug-out bag is intended for emergency survival during evacuations or disasters, typically providing enough supplies to last 72 hours. It includes essential items like food, water, first aid supplies, and shelter. A go-kit, on the other hand, is purpose-built for specific tasks, such as emergency communications, POTA, SOTA, or portable fieldwork. Go-kits can also serve as a backup station if your home station fails or you lose power, ensuring you can still communicate.

Emergencies and Radio Communications

Yes, a radio license is required, to operate a ham radio. However, in life and death emergency situations unlicensed operators may transmit on amateur frequencies. This is not to be the normal operating procedure, but something that, during a true emergency, could be necessary to save lives.

Baofeng radios have proven to be useful tools for staying in touch during disasters such as hurricanes, floods, and tornadoes. Their wide frequency range allows users to program local public safety, marine, and National Weather Service channels for listening (but not for transmitting), which further enhances the radios' usefulness in emergency situations.

The affordability and reliable performance of Baofeng radios make them a great choice for purchasing multiple units and storing them in the appropriate bag

Figure 4.1 — My bug-out and go-kit bags.

or waterproof case, regardless of whether they ever get pulled into service. They're also a convenient addition to the bug-out bags of licensed family members.

4.2 Necessary Bug-Out Bag Items

Whether for emergency communications, portable operation, or everyday carry use, bug-out bags contain some common components that are useful in every scenario. Let's start with your Baofeng radio of choice and the required items. The following sections outline some considerations for what to put in your bug-out bag(s).

In the next section, I will discuss some of the go-kits I use, but keep in mind that the bug-out bag will contain more items. Its role is more critical, and many of the components can also be used in a go-kit.

4.2.1 Power

Power sources are crucial in a bug-out bag because the 72-hour rule of thumb (mentioned above) suggests that you should plan and be prepared to go without power for several days. In situations like where the main AC power is unavailable, handheld radio users rely on battery packs and dc power sources to keep their devices charged.

Here are some suitable power options you can pack in advance in your bug-out bag to extend your handheld's battery life, as well as a few useful cable connectors. Remember, batteries require maintenance. They must be charged and stored properly to ensure that they provide maximum power output when you need them:

- ❒ Include one or two fully charged spare batteries. Consider an extended capacity battery for longer operation.
- ❒ A high-capacity power bank, ideally 20,000 mAh or more, compatible with USB port outputs and 12 V outputs.
- ❒ A USB charging cable for flexibility; allowing you to charge via power banks, car USB ports, or solar panels.
- ❒ External batteries having a 3 to 5 aH rating; common brands are Bioenno Power and Buddipole A123. These batteries are small, lightweight, and lithium-ion or LifePO4 chemistry, which is the preferred technology. Use these with a dc car adapter dummy battery pack.
- ❒ Portable 10 to 30-watt solar panels for off-grid charging of a power bank. They can also provide a 5 V USB output.
- ❒ A dc car adapter. This allows charging from a vehicle or power bank car accessory port.
- ❒ A dc battery eliminator. This replaces the battery with a dc regulator terminated on a car accessory plug.

You also must ensure that you have the appropriate power cables to connect to the radio. The Anderson Powerpole® has become the standard for radio power connectors, as it prevents reverse polarity and is easily customized. Baofeng does not provide an Anderson Powerpole® interface, but third-party vendors can make appropriate cables for you. In the past few years, all-in-one power stations like Goal Zero, Jackery, Bluetti, and EcoFlow have offered rechargeable power sources and can integrate with solar panels. These power sources are more for an emergency kit as opposed to an EDC kit, but smaller versions can be useful in go-kits.

4.2.2 Antennas

Depending on your situation, your antenna needs may vary. In almost every case, extending the range over which you can communicate is preferable to sticking with the Baofeng's stock, rubber duck antenna. Here's a list of antenna types to consider, along with a few notes about each type:

- A whip antenna, such as those from Signal Stick, Diamond, or Nagoya, are popular and improve range.
- A lightweight roll-up J-pole is another popular portable option for extending range. Signal Stuff also makes a collapsible aluminum rod J-pole antenna that fits in your backpack.
- A directional Yagi beam offers better performance and makes satellite operation possible, while also greatly extending your communication range. However, you may need a proper antenna tripod or mast for your Yagi deployment. Diamond, ELK Antennas, and Arrow manufacture Yagi antennas.
- A mobile magnetic mount antenna is ideal for vehicle use or temporary setups. (Remember, to realize any gain you must place the antenna on a metal object to form a ground plane.) Diamond and Comet offer many options, along with various antenna mounts for different vehicles.
- Fixed base antennas like the Comet GP-1 Dual-Band VHF/UHF Base Vertical Antenna are also an option. However, with a length of 50", these may require you to carry them in the trunk of a vehicle.
- Lightweight coaxial cables are necessary for connecting external antennas. RG-316 coaxial cable, with its compact and portable design, is ideal for this purpose when paired with SMA connectors. Baofeng radios use an SMA Female antenna connector, but not all antennas are compatible with it. Therefore, it's important to have a variety of adapters available to connect different types of antennas. Common antenna connectors include PL-259, SO-239, and BNC. These adapters are indispensable, especially when you can use someone else's antenna or need to operate from your vehicle.

4.2.3 Accessories

The following items are indispensable both in combination and individually:

- An external mic and speaker are low-cost items that increase the radio usability and clarity in outdoor or noisy environments.

- A headset or earpiece is useful for hands-free operation in tactical or noisy environments.
- A programming cable is essential for programming channels quickly in the field using a laptop or compatible mobile device. (For information on programming your Baofeng with a computer, see chapter 7.)

4.2.4 Supporting Documentation and Manuals

It's important to know how to use your equipment before you take it into the field. However, no matter how well you know your radios, there will always be moments when stress or forgotten details make it harder to recall the exact command sequence needed to achieve the desired result. Preprogramming your radio with frequencies for local repeaters, emergency services, weather channels, simplex, and any specific event frequencies will save you time out in the field and can help to lower your stress level. Even with proper preparation, it's a good idea to have copies of your equipment's documentation available when you need it. The following documents are always handy to have:

1. A printed-out a frequency list from *CHIRP*. Chapter 7 explains what *CHIRP* is and how to use it to program many frequencies into your radio at once.
2. A laminated cheat sheet of common operating procedures, frequency lists, and step-by-step guides for key functions (e.g., programming and setting tones).
3. A local repeater list. This should also be laminated or waterproof. The ARRL Repeater Directory is also a great resource.
4. The user manual for your radio. This should be a compact hard copy or a digital copy on a tablet or USB memory stick for quick reference.
5. A notebook and writing utensils for logging and notes.

I keep everything in waterproof, plastic pouches so that my paperwork stays dry and organized.

Figure 4.2 — Portable documentation examples. Store everything in waterproof baggies or containers

4.2.5 72-Hour Survival Items

The primary purpose of the bug-out bag is to help in surviving an emergency. This bag may be all you have to rely on, or it can act as a supplement to what is being provided by local emergency services. You should always feel free to customize the items as you see fit for special needs, but the foundation components remain the same. For example, one specific tip that I always use is helpful when packing medications. In most cases medications won't be put in your bag beforehand because you don't want them to expire or otherwise degrade, therefore I put an obvious tag on the top of my bag that reminds me to grab any medications I might need.

Here is a typical packing list of common items found in bug-out bags for your reference:

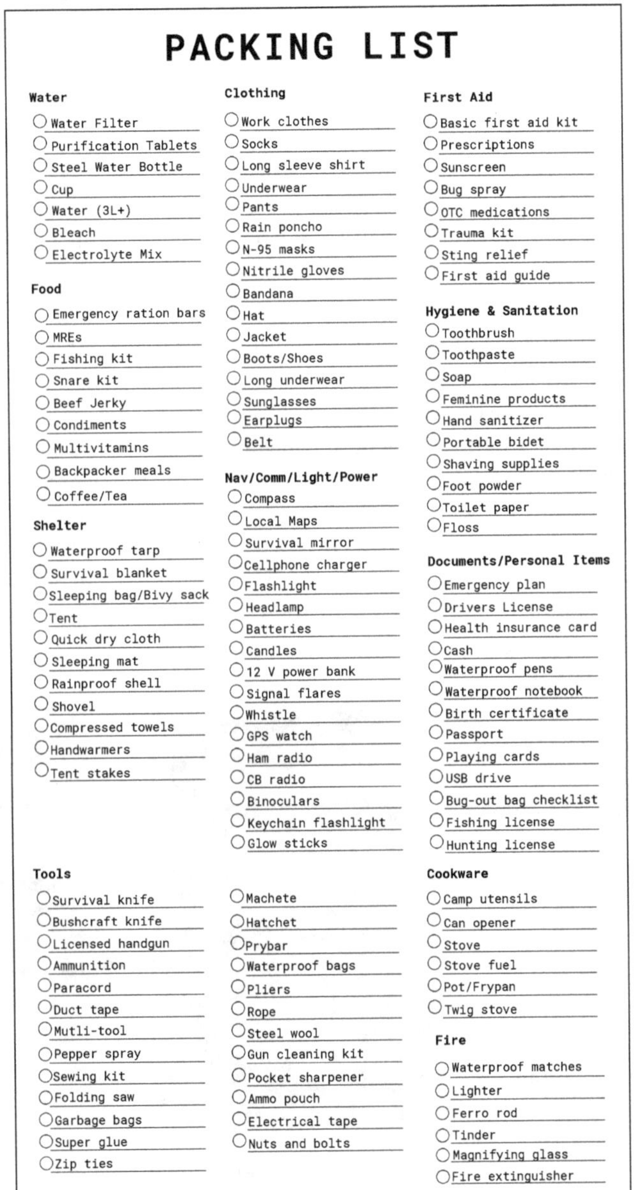

PACKING LIST

Water

- ○ Water Filter
- ○ Purification Tablets
- ○ Steel Water Bottle
- ○ Cup
- ○ Water (3L+)
- ○ Bleach
- ○ Electrolyte Mix

Food

- ○ Emergency ration bars
- ○ MREs
- ○ Fishing kit
- ○ Snare kit
- ○ Beef Jerky
- ○ Condiments
- ○ Multivitamins
- ○ Backpacker meals
- ○ Coffee/Tea

Shelter

- ○ Waterproof tarp
- ○ Survival blanket
- ○ Sleeping bag/Bivy sack
- ○ Tent
- ○ Quick dry cloth
- ○ Sleeping mat
- ○ Rainproof shell
- ○ Shovel
- ○ Compressed towels
- ○ Handwarmers
- ○ Tent stakes

Clothing

- ○ Work clothes
- ○ Socks
- ○ Long sleeve shirt
- ○ Underwear
- ○ Pants
- ○ Rain poncho
- ○ N-95 masks
- ○ Nitrile gloves
- ○ Bandana
- ○ Hat
- ○ Jacket
- ○ Boots/Shoes
- ○ Long underwear
- ○ Sunglasses
- ○ Earplugs
- ○ Belt

Nav/Comm/Light/Power

- ○ Compass
- ○ Local Maps
- ○ Survival mirror
- ○ Cellphone charger
- ○ Flashlight
- ○ Headlamp
- ○ Batteries
- ○ Candles
- ○ 12 V power bank
- ○ Signal flares
- ○ Whistle
- ○ GPS watch
- ○ Ham radio
- ○ CB radio
- ○ Binoculars
- ○ Keychain flashlight
- ○ Glow sticks

First Aid

- ○ Basic first aid kit
- ○ Prescriptions
- ○ Sunscreen
- ○ Bug spray
- ○ OTC medications
- ○ Trauma kit
- ○ Sting relief
- ○ First aid guide

Hygiene & Sanitation

- ○ Toothbrush
- ○ Toothpaste
- ○ Soap
- ○ Feminine products
- ○ Hand sanitizer
- ○ Portable bidet
- ○ Shaving supplies
- ○ Foot powder
- ○ Toilet paper
- ○ Floss

Documents/Personal Items

- ○ Emergency plan
- ○ Drivers License
- ○ Health insurance card
- ○ Cash
- ○ Waterproof pens
- ○ Waterproof notebook
- ○ Birth certificate
- ○ Passport
- ○ Playing cards
- ○ USB drive
- ○ Bug-out bag checklist
- ○ Fishing license
- ○ Hunting license

Tools

- ○ Survival knife
- ○ Bushcraft knife
- ○ Licensed handgun
- ○ Ammunition
- ○ Paracord
- ○ Duct tape
- ○ Mutli-tool
- ○ Pepper spray
- ○ Sewing kit
- ○ Folding saw
- ○ Garbage bags
- ○ Super glue
- ○ Zip ties
- ○ Machete
- ○ Hatchet
- ○ Prybar
- ○ Waterproof bags
- ○ Pliers
- ○ Rope
- ○ Steel wool
- ○ Gun cleaning kit
- ○ Pocket sharpener
- ○ Ammo pouch
- ○ Electrical tape
- ○ Nuts and bolts

Cookware

- ○ Camp utensils
- ○ Can opener
- ○ Stove
- ○ Stove fuel
- ○ Pot/Frypan
- ○ Twig stove

Fire

- ○ Waterproof matches
- ○ Lighter
- ○ Ferro rod
- ○ Tinder
- ○ Magnifying glass
- ○ Fire extinguisher

Figure 4.3 — A bug-out bag checklist.

4.2.6 Additional Items

There are several other tools that might be handy, and even necessary, to have in your bug-out bag. These include:

- ❒ A multi-tool (Leatherman, Victorinox, SOG) for repairs and adjustments.
- ❒ A flashlight or headlamp. You never know when you might have to operate at night.
- ❒ Electrical tape is great for quick patches and for binding objects together.
- ❒ Tie wraps make bundling cords quick and easy.
- ❒ At least 10' of 550 Paracord. A Navy Seal once told me to never travel without this, but even greater lengths of cordage of almost any kind are useful.
- ❒ Waterproof storage containers. These protect the radio and accessories in wet conditions. This can be the carry bag itself (backpack, duffle bag, etc.), but smaller, plastic storage containers are also useful to organize the contents of the bag.
- ❒ A small first aid kit is a must, and I recommend having a more robust first aid kit for emergency kit use.
- ❒ You may also choose to pack a laptop computer for digital modes, logging, or out in the field programming, depending on your use case. I use a Microsoft Surface tablet as well as an Apple iPad for various purposes. Today's smartphones are also robust computing devices, and an older, unused model is perfect for pressing into use as a backup unit. For example, my Google Pixel 4 does everything my current Google Pixel 6 does if I am connected to Wi-Fi. Download an SSTV app. They work great for sending images back and forth over the airwaves.
- ❒ A bag inventory check list. This allows me to do a quick 30-second inventory to make sure I haven't misplaced or forgotten anything. Before I started using my checklist, I often forgot one critical item that would prevent me from operating. A medication reminder, like I stated above, would also prove to be very useful.
- ❒ Pocket-sized digital voltmeter.

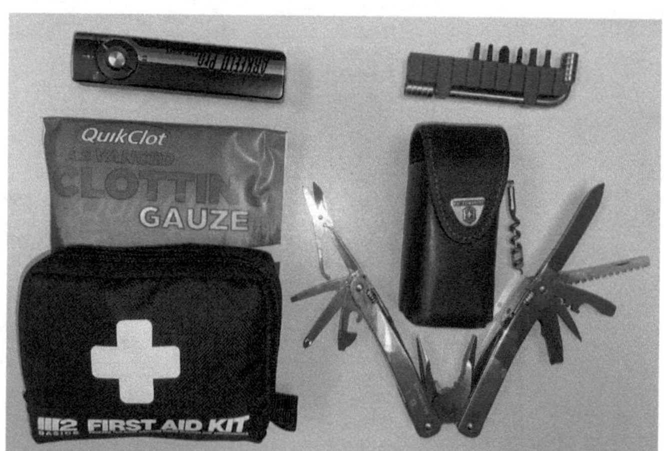

Figure 4.4 — Example bug-out bag items.

4.3 Necessary Go-Kit Items

If you're an active portable operator, public service volunteer, or just using your radio for every day carry use, go-kits share some common components that are useful in every scenario.

In the following sections, I will detail how I set up my travel go-kit and SOTA/POTA go-kits by using the list of bug-out bag components that I need for the specific operation task at hand. The affordability of accessories makes it easier to keep your kit fully equipped. For example, my frequently used kits each have their own speaker microphone and extra battery pack.

4.3.1 Travel Go Kit

I used to travel a lot for work, and during many of those trips I had a rental car. In those situations, I found that having a small and compact ham radio solution helped pass the time when driving several hours to customer locations. Ham radio is an ideal tool for communication with local hams while mobile or from hotel /motel rooms in the evening. I have a dedicated travel radio that has repeaters, simplex frequencies, NWS, and local public safety channels programmed in it for my travel destinations. Everything I need fits in a canvas military MOLLE pouch, packed in my suitcase ready to go, including:

- ❒ My radio
- ❒ Two Batteries
- ❒ A dc battery eliminator
- ❒ A mobile magnet mount antenna
- ❒ The Baofeng standard antenna
- ❒ A long gain Antenna
- ❒ A speaker/Mic

4.3.2 SOTA/POTA Go-Kit

For SOTA operations, I grab this go-kit along with my HF go-kit for urban activations (drive up locations). The only difference between this and my travel kit is that I add a Yagi antenna and coaxial cable. I have several antennas to use to help me complete my activation and provide the "hunters and chasers" with new location IDs to add to their logs.

If I was planning to do a SOTA activation on Mount Washington that requires hiking up to the summit, then my kit would include several survivals items from the bug-out bag as needed. You should always choose the components you need to bring specific to the task at hand. For example, parking in the Mount Diablo parking lot and walking a hundred feet to the summit would do just fine with the basic kit.

4.4 Continuous Preparation

Go-kits and bug-out bags should not be considered "make it and forget about it" packs. Be sure to inspect your electronic devices monthly. Also, perform monthly battery checks to ensure all the batteries in your kits are charged and ready for use. It would be even better to use the radio for a day or two, then charge and pack it again. In addition to being familiar with your equipment, and how to use and maintain it, you need to know how to deploy it quickly and easily. Before you need your bug-out bag or go-kit, deploy the antennas in them and check SWR readings to ensure they will work. Also, double-check that any required antenna adapters are packed.

If you are using multiple radios for maintaining communications in group, like for a family or Boy Scout troop, use *CHIRP* to program all the radios with the same channel assignments. When you tell someone to go to channel 67, you want to make sure everyone has the correct frequency programmed into channel 67.

Now let's discuss the phrase "One is none and two is one." This phrase contains a principle commonly cited in preparedness, survivalist, and operational planning circles. It emphasizes the importance of redundancy to ensure reliability in critical situations. Carrying a secondary handheld ensures you're not completely out of communication if your primary radio fails. Having both battery packs and a solar panel or power bank ensures you can stay operational even if one power source is depleted. A backup antenna can save the day if your primary antenna is damaged or doesn't perform as expected.

By embracing "One is none and two is one," you mitigate risks and enhance your ability to succeed under challenging circumstances. It's a simple yet powerful mindset for anyone who values preparedness and self-reliance.

I once had a situation where, after a one-hour hike up to a high spot for a simplex SOTA activation with a new Yagi antenna, I realized I had forgotten the coaxial cable, RF adapter, and the antenna instructions. The only usable antenna I had was the rubber duck that came with the radio. Since many of the stations I could hear were clearly at the edge of the antenna's normal coverage, the signals were weak and noisy, making it impossible to make contact with them. Two things that we've talked about in this chapter could have prevented this mistake. First, if I had used an inventory checklist before departing, then I would have caught that I had forgotten important equipment. On the other hand, even without a checklist, if I had also packed a high-gain whip antenna (which weighs a few ounces) I possibly would've still been able to complete my activation; this is an instance where following the "one is none and two is one" philosophy would've been useful.

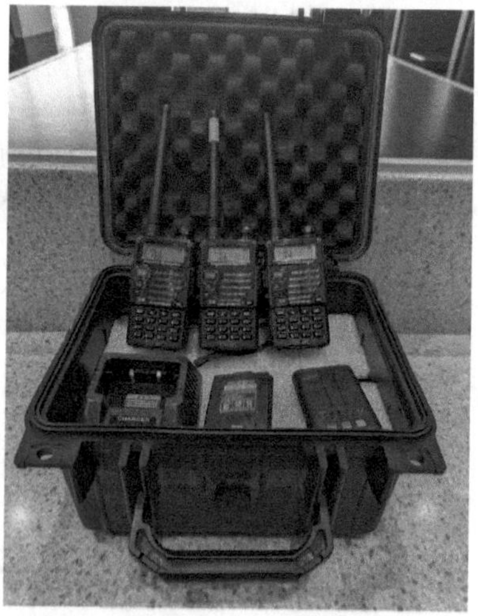

Figure 4.5 — A set of UV-5Rs programmed with channel assignments for use in a Boy Scout troop.

The key to a successful go-kit or bug-out bag is balancing portability, functionality, and reliability based on your specific needs and intended use cases, and — in some cases — the size of your wallet. If you use your kit regularly, you will add and delete items, constantly making the necessary changes to better meet your needs.

5 Emergency Preparedness and Satellites

5.1 Ham Radio in Emergencies

Ham radio plays a critical role in supporting agencies like FEMA, the Red Cross, and local emergency management organizations by providing reliable communication as a failsafe when primary systems are offline or overwhelmed. Because it operates independently of cell towers and the internet, ham radio is invaluable during natural disasters such as hurricanes, earthquakes, and wildfires, as well as during infrastructure failures.

The ham radio VHF/UHF bands are ideal for local communication, these bands often use ham repeaters powered by batteries or solar energy. Mobile stations, parked in strategic locations like fire and police departments or designated shelters, help facilitate real-time coordination for emergency services. These capabilities enable efficient resource allocation, supply distribution, and volunteer organization.

Figure 5.1 — Handhelds can be useful in a variety of emergency and public service settings.

Ham operators provide critical situational awareness by relaying real-time information such as weather conditions, road closures, and damage reports. They act as a communication bridge between disparate groups — like police, fire, and medical teams — that may use incompatible systems.

Ham operators are passionate about community service, often assisting with communications for marathons, parades, and search-and-rescue operations. Volunteers frequently use handheld radios to ensure smooth event management. For example, in bike races or endurance runs, operators monitor participants' safety, reporting progress and issues to event organizers. Another example of this is hams relaying ground-level weather observations to the National Weather Service (NWS) through programs like SKYWARN.

The Baofeng UV-5R is a popular choice in public service use because its low cost allows users to equip multiple family or team members with a radio communication device. It supports VHF and UHF bands, enabling access to local repeaters and simplex frequencies. The radios can be programmed for ham and public service frequencies (within legal limits). They offer useful versatility for ham communication, monitoring public service bands (receive-only), and basic scanning.

The most common limitation of the Baofeng UV-5R compared to professional-grade radios (e.g., Motorola, Vertex, Icom) is its build quality, which may impact reliability in harsh conditions. Its power output is limited to 4 to 5 watts, reducing range compared to higher-power mobile or base radios. The UV-5R's receiver is prone to interference in high-RF environments, such as near radio towers or dense urban areas.

While the Baofeng UV-5R is an affordable and versatile radio, its limitations make it less ideal compared to higher-quality equipment if its primary use is for demanding emergency and public service roles. However, with proper preparation and complementary gear, it is still a good radio and a valuable part of your emergency communication toolkit.

Figure 5.2 — Amateur radio operators and American Red Cross volunteers work together at net control during the Cascadia Rising emergency exercise.

When Disaster Strikes

Amateur radio can provide valuable assistance during times of crisis or disaster, when normal communication services overload or fail. Many people get into amateur radio specifically so they can be of assistance at those times.

ARRL, the national association for ham radio in the US, recognized the value of having a trained group of skilled operators who can be called upon during times of need, and in 1935 created the Amateur Radio Emergency Service® (ARES®), which "consists of licensed amateurs who have voluntarily registered their qualifications and equipment with their local ARES leadership, for communications duty in the public service when disaster strikes." Basic membership in ARES requires having an amateur radio license and a desire to serve, but that's just the beginning.

ARES members train continuously through online courses and in person drills to improve their communications and disaster response skills. Participating doesn't require lots of expensive gear, but knowing how to use what you've got is extremely important. Chapters 6 through 8 in this book will provide you with enough information to get up and running with your Baofeng. To learn more about ARES and how you can engage, visit **www.arrl.org/ares**.

5.2 Amateur Radio Satellites

Amateur radio has enabled communication via satellites, with the Radio Amateur Satellite Corporation (AMSAT) providing "birds in the sky" to facilitate global communication through satellite links. These satellites are basically radio repeaters taking a UHF as the input (uplink/transmit) frequency and outputting the signal to a VHF frequency (downlink/receive). The AO-91 and SO-50 (SaudiSat-1C) are common satellites that are used for ham radio. They require minimal equipment (a handheld with a dual-band whip or Yagi antenna) and allow single-channel operation.

Figure 5.3 — Demonstrating how to use the Arrow antenna for satellite work.

The SO-50 carries several experiments, including a mode J FM amateur repeater operating on 145.850 MHz uplink and 436.795 MHz downlink. The repeater is available to amateurs worldwide as power permits, using a 67.0 Hertz CTCSS tone on the uplink for on-demand activation.

The repeater consists of a miniature VHF receiver. The receive antenna is a 1/4 wave vertical mounted in the top corner of the spacecraft. The receive audio is filtered and conditioned then gated in the control electronics prior to feeding it to the 250 mW UHF transmitter. The downlink antenna is a 1/4 wave mounted in the bottom corner of the spacecraft and canted at 45 degrees inward.

Figure 5.4 — The Fuji-OSCAR99 cubeSat prior to launch. [Photo courtesy of the NEXUS Project Team]

The International Space Station (ISS) is another example of how satellites fit into the ham radio hobby. Astronauts aboard the ISS that hold ham radio licenses frequently communicate with students as part of an international outreach program called Amateur Radio on the International Space Station (ARISS). Many students learn about satellite communications through the program and get to talk with astronauts in space when the ISS passes over their school. Anyone with a VHF/UHF handheld radio can listen in by programming the frequencies 145.200 MHz for transmission and 145.800 MHz for reception. Aside from rare scheduled contacts with astronauts aboard

Figure 5.5 — Thor-Agena rockets carried several early amateur radio satellites into orbit.

the ISS, ham radio operators can make contacts with other hams on the ground using the ISS's powerful FM repeater. To access the repeater, operators must transmit on the 2 meter band at 145.990 MHz and listen on the 70 centimeter band at 437.800 MHz. Listening to the ISS as it passes overhead and making contacts with students during scheduled events is a truly joyful experience, as is progressing from just listening to making your own contacts via satellite. A popular activity is receiving Slow Scan TV (SSTV) images transmitted by the ISS, which can be decoded using smartphone or PC applications.

The Baofeng UV-5R and UV-82 handheld radios are commonly used by amateur radio enthusiasts for satellite communications, as they are dual-band radios. The UV-5R can handle simple satellite passes by manually switching frequencies. However, Baofeng radios (and others like them) have limitations for satellite operations. The biggest limitation is that neither radio supports "full duplex operation," meaning they can't transmit and receive simultaneously. Full duplex is ideal for satellite work because it allows you to hear your own transmission reflected back from the satellite. Additionally, Doppler shift affects the downlink frequency during the satellite's pass. Despite these challenges, satellite contacts can still be made by programming both the uplink and downlink frequencies into your radio's memory to accommodate the frequency shift.

Another ideal solution for "duplex" operation is to use two Baofeng radios, one for the uplink and the one for the downlink. Advanced operators find this method easier because two separate radios will feed two separate antennas. However, beginners can enjoy the satellites with just a single radio and a dual-band Yagi antenna.

It's important to note that,

Figure 5.6 — Using a simple dual-band, handheld transceiver and an Arrow handheld antenna can get you on FM satellites.

except for the repeater aboard the ISS, the standard antennas that come with Baofeng radios make FM satellite operations extremely difficult. Monitoring the satellite will only be possible during certain parts of its pass (when it's overhead). A longer 16-inch gain antenna can help, but for better performance, a dual-band (2 meter and 70 centimeter) Yagi antenna, with much higher gain, will significantly improve both transmit and receive signal strength during satellite passes. You can replace the factory antenna on Baofeng radios, as long as your cables have the correct adapters. For more information on Yagi antennas for satellite operations, refer to the Antenna section in Chapter 9.

Baofeng radios have allowed many new hams to quickly jump into the hobby and participate in activities that align with their interests. While some users are satisfied with the UV-5R alone, others choose to upgrade to different models and different brands as their interests grow. Better radios offer more opportunities for experimentation and enjoyment.

5.2.1 A Brief Guide to Practical Satellite Operation

The following is adapted from the June 2018 *QST* article by Sean Kutzko, KX9X, called "Get on the Satellites for ARRL Field Day."

More and more amateurs are discovering the fun of satellite operating. Stations can be small and portable. Satellite passes don't take much time, and even Technician-class hams have privileges on most satellites.

INTRO TO FM SATELLITES

Most of the FM satellites that are currently in orbit are known as CubeSats, a 10-centimeter cube with a dual-band FM transceiver and an antenna. They function like an orbiting repeater: only one station may transmit at a time. These satellites are known as low-Earth orbit (LEO). They are typically above the horizon for as little as a few seconds to a maximum of about 15 minutes. You can only communicate through a satellite if it is above your horizon. When a satellite first appears above the horizon at the beginning of a pass, that is called acquisition of signal (AOS); when it disappears below the horizon at the end of a pass, that is called loss of signal (LOS).

Most satellites transmit on one band and receive on another; the most common bands used are 2 meters and 70 centimeters. You transmit, or uplink, your signal to the satellite on one band, and receive, or downlink, the satellite's signal on the other band. At the time of this writing, a few FM satellites cover the US regularly, including SO-50 and AO-91. Most satellites make five to six passes over the US every 24 hours. While there are other satellites that use SSB or data, those are beyond the scope of this article.

Figure 5.7 —Many satellite operators use an extremely portable full-duplex station built around two Yaesu FT-818 transceivers for both FM and SSB (linear) satellites. [Katie Allen photo]

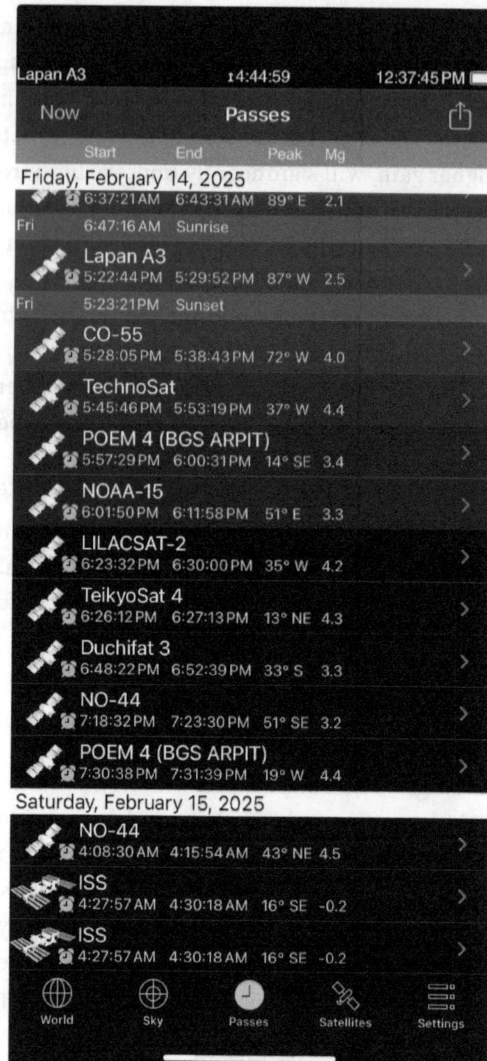

Figure 5.8 — Satellite passes for Newington CT, as shown in the GoSatWatch smartphone app for iOS.

THE GEAR YOU NEED

To be successful on satellites, you need to be able to transmit and receive on both 2 meters and 70 centimeters, antennas for each of those bands, and a way to track a satellite's progress over your part of the sky. For FM satellites, a dualband, handheld transceiver or two separate handhelds will work fine.

It is a big advantage to work satellites in full-duplex mode; this means you can hear the satellite's downlink as you are transmitting, allowing you to know when your signal is making it into the satellite. You can also operate satellites in half-duplex mode, meaning that when you transmit, you can't hear the satellite's signal, but this puts you at a disadvantage. Being able to know when you are making it into the satellite is a huge factor in a successful contact. The FM satellites listed above require a 67.0 Hz CTCSS tone on your transmitted signal.

It's possible to make satellite contacts using simple vertical antennas, but you'll have more success if you use directional antennas with gain, such as a small Yagi, for both bands.

It's easy to build your own antenna, or there are several commercial antenna options for satellites, such as the Arrow or Elk antennas, which can be held in one hand and easily follow the satellite as it passes overhead.

You will need a diplexer, which isolates your transmitted signal from your receive radio. Without it, you will likely overload the front end of your receive radio and make contacts impossible. Several manufacturers offer diplexers; the Arrow antenna has an optional diplexer that is stored in the antenna handle.

Lastly, you'll need to know when satellites will be overhead and where to point your antenna. Resources that offer tracking information include the *GoSatWatch* ($10) smartphone app for iOS users, the *AMSATDroid* app for Android users, AMSAT's website, and **n2yo.com**.

ADJUSTING FOR DOPPLER

The Doppler effect impacts satellite operating and is more noticeable on 70 centimeters than on 2 meters; as a satellite pass progresses, you will need to adjust your 70-centimeter frequency. Depending on the satellite, this will be either your transmit or receive frequency. If you don't adjust for Doppler, you will either not properly hear the downlink, or you will not be transmitting on the correct frequency to get into the satellite. See the sidebar, "FM Satellite Frequency Chart," for proper frequency adjustments for the four FM satellites. You can program these into your radio and adjust for Doppler easily.

STEPS TO SATELLITE SUCCESS

As with any new undertaking, don't expect instant success. Learning how to make a contact through a satellite, especially during hypercrowded conditions, will depend on your abilities. There are some concrete things you can do to maximize success:

1. Practice. Do not expect to simply turn on the radio and make a satellite contact. Take plenty of time to learn how to track satellites, aim the antenna throughout a pass, and listen for stations.
2. If you know a satellite operator, ask them to mentor you. Their expertise will only help you. You can also join the AMSAT Facebook group and get pointers from thousands of satellite operators who want to help you succeed.
3. Read the "FM Satellites Best Practices Guide" on AMSAT's website at **www.amsat.org/fm-satellites-good-operating-practices-for-beginning-and-experienced-operators/**. There are lots of good tips there.
4. If you live near the coasts, try working a pass that is mostly over water. This will reduce the number of stations that will have access to the satellite. During the extremely crowded conditions, this will work in your favor, and increase the odds of you making a contact.
5. Consider passes in the middle of the night. Fewer people will be on, which will increase your chances of success.

6. Remember, FM satellites are repeaters in the sky. Only one person can talk on them at a time. Make your contact and let others make theirs.
7. Don't transmit if you cannot hear the satellite. Even though you can't hear, you may still be getting into the satellite, which will interfere with other stations' contacts. This is unacceptable. If you are trying to work a satellite and do not hear yourself or other stations, stop transmitting and review your equipment; you may have a problem.
8. Your operating site should have as clear a view of the horizon as possible. Buildings and trees can interfere with your ability to hear the satellite; an open clearing on a high spot with no obstructions is the ideal operating location.
9. Know what Maidenhead grid square you are operating from. While it's not required, many satellite operators exchange grid information during contacts.
10. If you have the capability, try SSB satellites instead. There is more of a learning curve, but with 20 KHz of bandwidth or more, SSB satellites can facilitate multiple contacts simultaneously, like being on an HF band. More information is on the AMSAT website.

Working the FM satellites is a lot of fun. With a little practice and education, you can increase your odds of success and continue in this exciting area of amateur radio.

FM Satellite Frequency Chart

Here's a list of frequencies used for five different phases of a satellite pass, from Acquisition of Signal (AOS) to Loss of Signal (LOS). These changes will help adjust for the Doppler effect.

SO-50 (2-meter uplink/ 70-centimeter downlink)

Part of Pass	*TX Freq. (67Hz CTCSS)*	*RX Freq.*
AOS	145.850 MHz	436.805 MHz
	145.850 MHz	436.800 MHz
Middle	145.850 MHz	436.795 MHz
	145.850 MHz	436.790 MHz
LOS	145.850 MHz	436.785 MHz

AO-91 RadFXSat/Fox-1B (70-centimeter uplink, 2-meter downlink)

Part of Pass	*TX Freq. (67Hz CTCSS)*	*RX Freq.*
AOS	435.240 MHz	145.960 MHz
	435.245 MHz	145.960 MHz
Middle	435.250 MHz	145.960 MHz
	435.255 MHz	145.960 MHz
LOS	435.260 MHz	145.960 MHz

6 Manually Programming the UV-5R

6.1 UV-5R Overview

I recommend purchasing your radio from an authorized retailer to ensure you get a unit that's under warranty. Note that some bundles may vary slightly depending on the vendor or specific model variant, but the standard box contains the following items.

- **Baofeng UV-5R** – The handheld transceiver.
- **Antenna** – A flexible 5" dual-band antenna (VHF/UHF) with an SMA-Female connector.

- **Battery pack** – A rechargeable Li-on battery, typically rated at 1800 mAh.
- **Belt clip** – A screw-on clip for attaching the radio to your belt or pocket.
- **Lanyard** – A wrist strap for additional portability.
- **Desktop charger** – A cradle-style charger with ac adapter for charging the battery.
- **Earpiece with inline microphone** – A simple wired earpiece with a push-to-talk (PTT) button and microphone.
- **User manual** – A basic instruction manual that explains the radio's features, programming, and operation.

Note: If you want to program your radio using a computer (which I highly recommend and detail in Chapter 7), you need to purchase the optional K-Plug Programming Cable or a KY4-USB from RT Systems or something similar. Please note that there are many cheap cables out there with inconsistent chipsets that may or may not work. In this case, it's best to buy a name brand.

The basic specifications of the Baofeng UV-5R follow.

- A frequency range of 136 – 174 MHz (VHF), and 400 – 480 MHz (UHF)
- A power output of 1 W / 4 W selectable
- 128 programmable channels
- An 1800 mAh Li-ion rechargeable battery (7.4 V)
- An SMA-male antenna connector
- A dual-band display
- Dimensions of 4.33 inches × 2.36 inches × 1.26 inches
- Weight of approximately 7.5 ounces (including battery)
- A 2-pin Kenwood-compatible audio jack
- Additional features: FM radio, flashlight, voice prompts, and scan functions

6.2 Basic Use and Getting Started

Baofeng radios have several features that are not commonly found in other radios, such as their built-in LED flashlight for low-light environments, variable color backlights, and alarms. The alarm sends a loud alert signal, which can be useful as a safety feature, but needs to be used properly so that it doesn't cause unwanted and illegal radio interference. Additionally, the radio can tune into FM broadcasts, allowing you to listen to stations for entertainment or news when you're not actively using the radio.

When you first get your radio, you should do the following:

- ❒ **Unbox.** It may go without saying, but you'll need to unbox everything and remove the various components from their plastic wrapping.
- ❒ **Attach the antenna and battery pack.** Screw the antenna on (hand tight) and then slide the battery pack into place.
- ❒ **Attach belt clip:** Using the supplied screws, attach the belt clip to the battery pack. You'll need a computer or electronics-sized cross driver (small).
- ❒ **Charge the battery.** Unlike higher end electronic equipment, Baofeng radios do not come with charged batteries, so right away you'll need to plug in the charger, slip the radio into it, and give the battery a charge. Note that the battery has two slots that fit over two ridges in the charger to help secure the radio. For the 1800 mAh BL-5 battery, it takes about 4 to 5 hours to fully charge. If you're using the 3800 mAh BL-5L battery, it will take approximately 7 to 8 hours for a full charge.

  The battery charge indicator will show a solid red light while charging and turn solid green once the battery is fully charged. If the indicator flashes red and green, it may indicate a faulty battery pack.

  Please be aware that a Li-ion battery reaches its full strength only after being depleted and recharged at least three times. After these three cycles, you can charge the battery at any time without worrying about what's commonly called memory effect. Memory effect is when a battery discharges or loses power more quickly than it should. I won't go into a technical discussion, but just know that to get the longest life from your battery, you need to charge and discharge it three times in a row.

In addition, Li-ion batteries can overheat and start fires when overcharged, so don't leave the radio in the charger for long periods of time when the battery is fully charged. The battery and/or the charger should have an overcharge circuit protector that prevents overcharging, but in case that's faulty, it's best to be careful.

- ❐ **Turn on the radio.** Turn the radio on. If you'd like, you can push the orange CALL button (above the large PTT button) to listen to a local FM broadcast station while I walk you through programming the radio. Some radios come with the English voice activated, and it announces the menu items and digits being pressed. However, if you reset the radio to factory defaults, it reverts to Chinese. This feature is very helpful for navigating the menu.

Variable Frequency Oscillators (VFOs)

The circuit that really makes a radio is the oscillator. An oscillator takes a direct current (dc) source and produces and alternating current (ac) signal, represented as a sine wave. A sine wave has both an amplitude and a period (take a look at the figure below). The amplitude is how high and low the peaks of the wave reach, and the period is how long it takes the wave to move through one cycle. Period and frequency are inversely related. As the frequency of a signal goes up, the time it takes to complete one cycle goes down. The tone from a tuning fork is one example of a sine wave. So is a radio emission. The simplest oscillators produce and receive only a single frequency. If you have an oscillator that can be tuned to multiple frequencies (think about "tuning" the radio in your car), you have what's called a variable-frequency oscillator. That means it can be made to resonate at a variety of frequencies. Since your Baofeng radio receives and transmits over a range of frequencies, it has a variable frequency oscillator within it, which is frequently abbreviated to VFO. Having multiple VFOs allows a radio to tune across multiple sets of frequencies simultaneously.

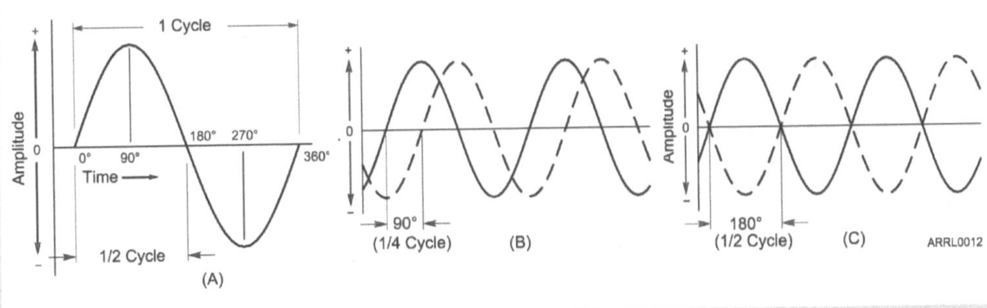

6.3 UV-5R Walkthrough

The following section provides a button-by-button walkthrough of the user interface on the Baofeng UV-5R, including all of the buttons and the LCD display.

6.3.1 Front Panel

Display

The UV-5R has an informative two-line LCD display that shows key information such as the frequency, battery level, signal strength, mode indicators, power level, and any activated features (like CTCSS tones). The display backlight can be set to different colors for various modes.

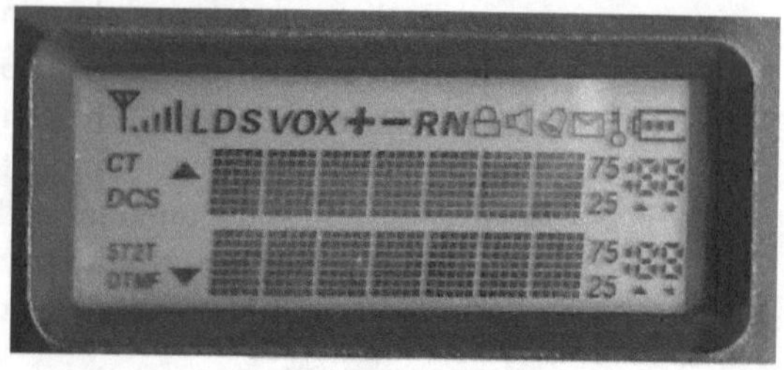

Below, I've provided an explanation of the on-screen icons and indicators. The explanation starts at the top left and works its way down through the three display lines,

Antenna Icon – indicates the transmitted and received signal strength in much the same way a cell phone does. More bars indicate a stronger signal.

L – Indicates the power level is set to low. When set to high, the L goes away.

D – Undefined.

S – indicates that the Dual Watch/Dual Receive function is activated. The S goes away when the function is deactivated.

VOX – is an abbreviation meaning Voice Operation. It's a hands-free PTT function. When activated, VOX displays. VOX goes away when deactivated.

+- – Indicates the direction of the offset, whether for a repeater or for split use.

R – Indicates that the reverse function is activated. Operating in reverse means listening on the input frequency and transmitting on the output. This feature let's you check if you're within simplex range of users who may be weak on a repeater. Not usually used for transmitting.

N – indicates that the Narrow FM mode has been activated. Ham radio uses wide FM (25 kHz) in general. FRS and GMRS often use narrow FM (12.5 kHz) in order to fit more channels in a given bandwidth.

Lock Icon – Undefined.

Speaker Icon – Undefined.

Bell Icon – Undefined.

Envelope Icon – Undefined.

Key Icon – Indicates the keypad is locked. The key icon goes away when the function is deactivated.

Battery Icon – Indicates the current battery level. More bars indicate longer battery life.

CT – Indicates CTCSS tones are activated. CT goes away when the function is deactivated. See Table 3.1 for a list of CTCSS tones.

DCS – Indicates digital coded squelch is activated. DCS goes away when the function is deactivated. Digital coded squelch is similar to CTCSS, but with 100 possible codes that use slightly more bandwidth. See **Table 6.1**.

▲ – Indicates VFO A is in use.

Frequency A – Indicates the frequency VFO A is set to. Note that the Baofeng UV-5R is not a true dual VFO radio. Instead of using two VFOs the radio rapidly switches between listening on two frequencies.

75/25 – Indicates the operating frequency.

188 – Indicates the operating channel.

5T2T – Undefined.

DTMF – Indicates dual tone multi-frequency (DTMF) is enabled. DTMF tones are the sounds emitted by a touchtone phone. They correspond to the numbers 0 – 9. Each number is indicated by two tones (one high and one low) that are transmitted on the voice channel. For a fuller discussion of DTMF tones and uses, see the DTMF Settings section of chapter 7.

▼ – Indicates VFO B is in use.

Frequency B – Indicates the frequency VFO B is set to. Note that the Baofeng UV-5R is not a true dual VFO radio. Instead of using two VFOs the radio rapidly switches between listening on two frequencies.

75/25 – Indicates the operating frequency.

188 – Indicates the operating channel.

Table 6.1

DCS (DTCS) Numbers and Codes

No.	Code	No.	Code	No.	Code	No.	Code	No.	Code	No.	Code
1	023	15	114	29	174	43	315	57	445	71	631
2	025	16	115	30	205	44	331	58	464	72	632
3	026	17	116	31	223	45	343	59	465	73	654
4	031	18	125	32	226	46	346	60	466	74	662
5	032	19	131	33	243	47	351	61	503	75	664
6	043	20	132	34	244	48	364	62	506	76	703
7	047	21	134	35	245	49	365	63	516	77	712
8	051	22	143	36	251	50	371	64	532	78	723
9	054	23	152	37	261	51	411	65	546	79	731
10	065	24	155	38	263	52	412	66	565	80	732
11	071	25	156	39	265	53	413	67	606	81	734
12	072	26	162	40	271	54	423	68	612	82	743
13	073	27	165	41	306	55	431	69	624	83	754
14	074	28	172	42	311	56	432	70	627		

Keypad

The keypad consists of a 16-button grid that includes numbers and several multifunction keys. Each key serves various purposes, such as for navigation, menu settings, and direct frequency input.

VFO/MR – This button, positioned on the front top left side of the radio, toggles between VFO (Variable Frequency Oscillator) mode for direct frequency entry and MR (Memory Recall, which is just the radio's memory). I'll show you how to program memory channels in section 6.4.3.

A/B – This button toggles between the A/B VFOs (top and bottom displays) in dual-watch mode.

BAND – This button switches between VHF (136 to 174 MHz) and UHF (400 to 480 MHz) frequency bands in VFO mode.

MENU – This button opens the settings menu where you can adjust frequencies, tones, power levels, and other settings.

▲/▼ – Use the up and down arrow keys to scroll through the different menu options.

EXIT – The exit button exits the current menu or setting or cancels an operation.

LED Indicator – A small LED light below the VFO/MR button glows different colors to indicate the radio's status. If the light is red that indicates that the radio is transmitting, if the light is green then the radio is receiving a signal.

When the **MENU** key is pressed, the radio will verbally say "Menu." This key also enters whatever option modification you make to the radio. For example, after modifying a menu option, you should press the **MENU** key again to accept the change. The radio will say "Confirm" to acknowledge a change in the menu's parameters. The menu will stay active for around 10 seconds after changing a menu option before timing out.

Each number key has dual functions depending on whether the radio is in standby or menu mode. Numbers 0 through 9 are used to enter frequencies directly in VFO mode, select menu options by number in the menu, and used for direct channel selection in memory mode.

Keypad Shortcut Keys

Table 6.2 provides a key for the different shortcut menu options and functions that the numeric keys on the UV-5R provide. In addition to the numeric keys, the keypad on the radio also has a ***SCAN** key that initiates a scanning mode to search for active frequencies, and a # key that locks the keyboard.

Table 6.2

Menu Shortcut Keys

Key	*Shortcut Menu Option*	*Function*
1	STEP	Sets the frequency step size.
2	TXP	Sets the transmit power (HIGH/LOW).
3	SAVE	Toggles battery saver mode.
4	VOX	Toggles VOX (voice-operated transmission).
5	WN	Sets the bandwidth (Wide/Narrow).
6	ABR	Adjusts display backlight timeout.
7	TDR	Enables/disables dual watch (monitoring two frequencies).
8	BEEP	Keypad beep
9	TOT	Transmission times out after a certain time
0	SQL	Adjusts the squelch level.

6.3.2 Left Side Buttons

The left side of the UV-5R is home to three important buttons, the **PTT (Push-to-Talk)**, the **CALL** button, and the **MONI** button.

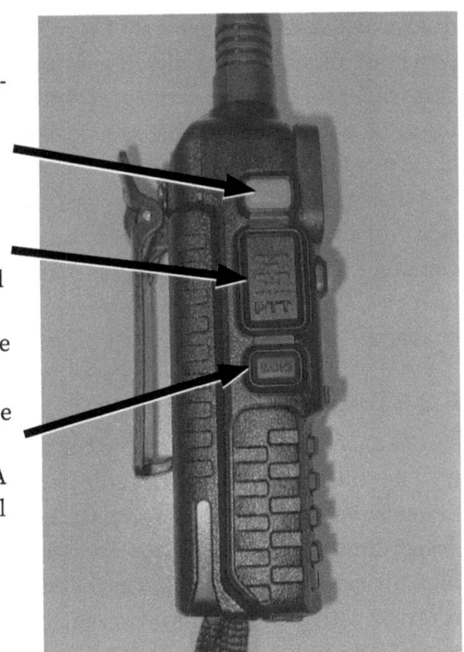

CALL – The orange CALL button is positioned above the PTT. When pressed, it will toggle between the VHF/UHF channel and the built-in FM radio meant for listening to commercial FM broadcasts.

PTT (Push-to-Talk) – This button activates transmission when pressed, enabling voice communication. When you select your desired frequency, indicated by the highlighted display row, press down the PTT button to transmit and then release to return to receive mode when finished.

MONI – This button, positioned underneath the PTT, serves two different functions depending on the length of time you press press it. A long press opens the squelch, which is useful for weak signals. A quick press toggles the built-in flashlight function.

6.3.3 Top Controls

The top of the UV-5R has three important features, including the **Antenna Connector**, the **Volume Knob/Power Switch**, and the **Flashlight**.

Antenna Connector: The UV-5R uses a standard SMA-male port. This allows users to swap the stock antenna with compatible aftermarket antennas for better performance. The antenna has a SMA-female connector.

Volume Knob/Power Switch: The top rotary knob controls the radio's volume and serves as the power switch. Turning it clockwise powers the radio on. Rotating farther increases the volume, and rotating it counterclockwise powers the radio off.

Flashlight: The UV-5R includes a small LED flashlight, activated by a quick press on the MONI button. This function can be helpful in low-light conditions.

6.3.4 Right Side Speaker/Mic Port:

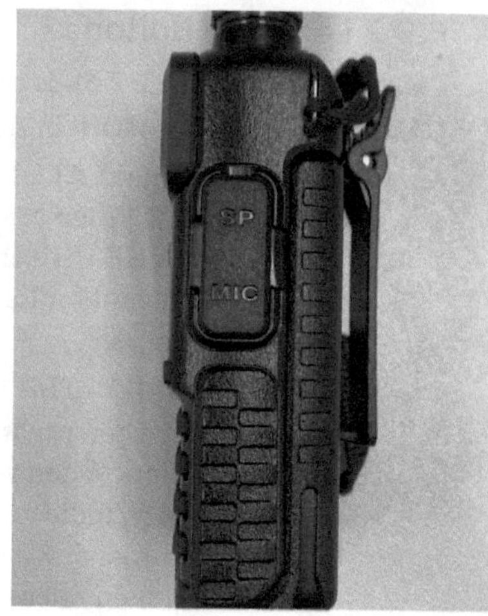

On the right side of the UV-5R is a dual-pin port that allows connection to external accessories such as a speaker, mic, or programming cable. This port uses the Kenwood radio standard, which is an arrangement of speaker and microphone ports that allows for easy digital programming. The measures differ for each pin in the port, with the top pin measuring at 3.5 mm for the speaker, and the bottom pin measuring at 2.5 mm for the microphone.

When programming with software like *CHIRP,* use the side port and a compatible cable to connect the radio to a computer. This is the easiest way to manage channels, frequencies, and other settings. I cover this process in the next chapter.

6.3.5 Back Panel

The only notable feature on the back panel of the UV-5R is the battery pack. The radio uses a removable Li-ion battery pack that locks onto the back panel. This design allows for easy replacement or upgrading to a higher-capacity battery pack. The belt clip, which you screw on with provided hardware, can be easily removed if desired.

Menu List

A list of the most commonly used UV-5R menu items appears at the end of this chapter in section 6.7 Settings Quick Start Guide. Section 6.11 Cheat Sheet is also extremely handy.

6.4 Manual Programming

There are several programming methods for the UV-5R, including quick programming with manual entry, simplex memory programming, and programming repeaters. In this section, I'll walk you through each of these methods step-by-step, with a detailed button-by-button guide.

Before you begin, ensure that your radio is properly set up for optimal performance. First, check that you have a fully charged battery. You'll want to ensure that the battery can last through the programming exercise. Next, you'll choose a channel, for this example I chose a local NWS 24/7 weather channel. In my area, Environment Canada uses the 162.400 MHz frequency for weather reporting. When listening to this frequency, you will receive the local weather broadcast for your area. If you do not clearly receive the broadcast, you can tune between 162.400 to 162.550 MHz until the broadcast comes through more clearly. (The National Oceanic and Atmospheric Administration [NOAA] in the United States broadcasts weather forecasts using the same frequencies, and its coverage is basically ubiquitous.) When choosing a channel, I strongly recommend enabling the low power setting in the event you mistakenly transmit, doing so will limit the range of any transmission and help avoid interference. This setting can also be used for a quick frequency check before deciding whether to put it into memory. I will walk you through changing to low power in the below section. A female voice will announce the button keypresses as we walk through the steps so long as the default "voice prompts" is enabled.

6.4.1 Programming the Voice Menu on the UV-5R

Before getting into the programming methods, let's ensure that the voice prompts are enabled and in a language you understand. The default language for the UV-5R is Chinese, however you can choose to have it speak English by adjusting a menu option. Below is a step-by-step guide to changing the language of the voice prompt from the default to English.

1. Turn on your radio by rotating the **Volume Knob/Power Switch** clockwise until the LCD display turns on.
2. Press **MENU** to open the menu options (see figure).

3. Navigate to option 14 using the ▲/▼ buttons (see **Figure 6.8**) or by entering "14" on the numerical keypad. The radio's display screen will read **VOICE CHI** to indicate that the default voice setting is in Chinese. Then press **MENU** to confirm, at which point the radio will confirm, in Chinese, that you are on the correct menu setting for voice prompts.
4. Using the ▲/▼ buttons, navigate to the **ENG** option, which will show on the display screen, then press **MENU** to confirm the choice. When this is completed correctly the radio will announce "Confirm" in English.
5. Press **EXIT** to return to the main frequency display (see middle figure).

When navigating through the menu options try to be swift. The sub-menu does time out after a brief period, bringing you back to the main frequency display. However, you can always navigate back to the menu if the timer runs out on you by pressing **MENU**.

Eventually, when you feel comfortable enough with your radio, you can change the voice setting to **OFF** in the same sub-menu setting using the same method.

6.4.2 Manually Enter a Quick Frequency

Now that we have the voice prompt enabled, I will detail a step-by-step method for manually entering a quick frequency into the UV-5R.

1. Press **VFO/MR** to switch to VFO mode until the display shows a frequency rather than a channel name or number (see bottom figure). When you are on the correct display the radio will announce "Frequency Mode" through the speaker. If you do not hear this, adjust the volume knob and try again.

2. Press **A/B** to select the upper display. An upward-facing arrow indicates the selected frequency.
3. Use the numerical keypad to enter the desired frequency. In my example, that would be: 1 6 2 4 0 0 for 162.400 MHz. The radio will announce the numbers as they are being entered into the keypad.
4. Press **EXIT** to confirm the entered frequency. If you are following along with my example, you should now be able to receive the local weather reports on 162.400 MHz.

Adjusting the Power Level

Before moving on to the next method, I will walk you through changing the power level to low power so that if you accidentally transmit while setting the radio up, your transmission is as limited as possible.

1. Press **MENU**. The radio announces "Menu," and the display screen shows the "0" menu option default, which reads **SQL 5**.
2. Navigate from menu option **0 SQL** to **2 TXP** using the ▲/▼ buttons or by pressing the **2 TXP** button on the front panel. Then press **MENU** once more. At this point, the radio will announce "Power" to indicate that you are in the power settings.
3. Use the ▲/▼ buttons to navigate to the low power option. The display should read **TXP LOW** at this point. Then press **MENU** once more to confirm, the radio will announce "Confirm" once this is done.
4. Press **EXIT** to return to the main frequency display screen.

You can now enjoy listening to your local weather broadcast

Below you can follow along with the step-by-step process to manually program three simplex frequencies into memory on the Baofeng UV-5R, including: Memory 001: 146.520 Simplex, Memory 002: 446.000 Simplex, and Memory 003: 162.400 Weather Channel.

Enter a Simplex Frequency

1. Press **VFO/MR** to enter VFO mode. The radio will also announce "Frequency Mode." If it says "Channel Mode," you need to hit the VFO/MR button one more time.
2. Press **A/B** to select the upper display. The **A** (or upper) option is used to program the repeater channels into the radio. Programming data entered on the **B** (or lower) option will not be saved.
3. Use the numerical keypad to enter the desired frequency **1 4 6 5 2 0** for 146.520 MHz. The radio will announce the numbers as they are entered.

Turn Off Repeater Shift

Our next step is to turn off the repeater shift, as we are using a simplex frequency and don't need it turned on.

1. Press **MENU.**
2. Navigate to menu option **25 SFT-D** using the ▲/▼ buttons, or by entering **25** on the numerical keypad, then press **MENU** to confirm your selection. The radio will announce "Frequency Direction" when you've finished this step.
3. Using the ▲/▼ buttons, toggle to **OFF** and then press **MENU**, at which point the radio will announce "Confirm" and you can press **EXIT** to get back to the main frequency display.

Set the Transmit Power Level

Now we can set the transmit power level.

1. Press **MENU.**
2. Navigate to menu option **2 TXP** using the ▲/▼ buttons, or by entering **2** on the numerical keypad, then press **MENU** to confirm your selection. The radio will announce "Power" when you've finished this step.
3. You can now select either **HIGH** or **LOW** using the ▲/▼ buttons, then press **MENU** to confirm, at which point the radio will announce "Confirm" and you can press **EXIT.**

Turn Off Dual Watch/Dual Standby

We now need to turn off TDR *(Dual Watch/Dual Standby)* which allows the radio to monitor both VFO A and VFO B at once.

1. Press **MENU.**
2. Navigate to menu option **7 (TDR)** using the ▲/▼ buttons, or by entering 7 on the numerical keypad, then press **MENU** to confirm your selection. The radio will announce "Dual Standby" when you've finished this step.
3. Using the ▲/▼ buttons, toggle to **OFF**, then press **MENU** once more to confirm, which the radio will announce, and **EXIT** to return to the main frequency display.

Delete Memory Channel 001

Now we will clear the target memory 001 before we enter the frequency into a memory channel.

1. Press **MENU.**
2. Navigate to menu option **28 (DEL-CH)** using the ▲/▼ buttons, or by entering **28** on the numerical keypad, then press **MENU** to confirm your selection. The radio will announce "Delete Channel" when you've finished this step.
3. Use the ▲/▼ buttons or enter **001** on the numerical keypad. The display screen will show **DEL-CH 001** when you have entered everything correctly.
4. Press **MENU** to confirm the selection, then **EXIT.**

Save Memory Channel 001

Now we can enter the VFO frequency into channel number **001.**

1. Press **MENU.**
2. Navigate to menu option **27 (MEM-CH)** using the ▲/▼ buttons, or by entering 27 on the numerical keypad, then press **MENU** to confirm your selection. The radio will announce "Memory Channel" when you've finished this step.
3. Use the ▲/▼ buttons or enter **001** on the numerical keypad. The radio will display **MEM-CH CH-001** when you have entered everything correctly.
4. Press **MENU** to confirm. The radio will announce "Receiving Memory."
5. Press **EXIT.**

To program the transmitting memory, follow the steps above to navigate to **CH-001.** This time, when you press **MENU** to confirm your selection the radio should announce "Transmitting Memory."the 2 meter national calling simplex frequency is now stored in channel 001.

For programming the other two simplex frequencies into memory on the Baofeng UV-5R noted above, 446.00 Simplex on Memory Channel 002 and 162.400 Weather Channel on Memory Channel 003, you will enter the new frequencies one at a time in VFO A (446.000 and then

162.400), and follow the steps for Delete Memory Channel 001 and Save Memory Channel 001, assuming that you want to duplicate the other settings for each new channel.

Verifying your Programming

Now let's verify that your programming of the three simplex frequencies into memory is correct. To do this while in VFO mode, press **VFO A/B**. The radio will announce "Channel Mode." Then use the ▲/▼ buttons to look through the channels. You can also enter the 3-digit channel number that you specifically want to see. The display screen will show you what frequency is stored in that memory, and the radio will announce the channel numbers as you cycle through the memories.

We have now programmed several memory channels into the following memory channel locations:

Memory 001: 146.520 Simplex

Memory 002: 446.000 Simplex

Memory 003: 162.400 weather channel

Let's make a first contact with a ham friend to test out the radio. You can make contacts on channels 1 and 2, remember that channel 3 is for listening only. On the UV-5R, channels cannot be labeled directly on the radio. If you need names, consider using programming software like *CHIRP*. This will be covered in Chapter 7.

6.5 How to Manually Program a Repeater Channel

Let's start with programming your local radio club repeater. Gather the frequencies, offset, and tone settings (CTCSS) of the repeater from your repeater directory resource of choice.

Before we start programming it might be useful to revisit chapter 3, which discusses repeater operation at a high level and provides detailed explanations of offset and CTCSS.

6.5.1 Repeater Offset

Repeaters allow a signal that appears on the input frequency to be retransmitted on the output frequency. There are standard offsets (splits) assigned and for VHF it's plus or minus 600 KHz and for UHF it's plus or minus 5 MHz.

6.5.2 CTSS/DCS Codes

Continuous Tone-Coded Squelch System (CTCSS), is a feature that helps radios selectively receive signals by using a sub-audible tone. This tone is transmitted along with the main audio signal and can open or close the squelch (muting) of a receiving radio. It's commonly used on repeaters and shared frequencies to filter unwanted transmissions, reducing interference and ensuring that users only hear relevant transmissions. DCS or Digital Code Squelch is a digital version of CTCSS, and it's not often used on ham repeaters in my experience.

How CTCSS Works

You can set up your Baofeng to send a CTCSS tone or to open up its squelch when one is received.

- **Transmit CTCSS Tone** – Your radio transmits a continuous sub-audible tone at a set frequency (e.g., 100 Hz) whenever you press PTT. Only radios with the same CTCSS tone configured on the receiving end will hear your transmission. Most repeaters will just require this for the input frequency.
- **Receive CTCSS Tone** – Your UV-5R will remain silent unless it detects a transmission with the exact CTCSS tone you've set. This prevents interference from other users who may be using the same frequency but a different tone. The repeater frequency information will indicate if this is required.

For more on CTCSS tones, see section 3.2 Continuous Tone-Coded Squelch System in chapter 3 Repeaters. For information on programming the CTCSS tone for transmit and/or receive, see the section in chapter 7 on Programming Memory Entries (step 8).

6.5.3 Programming a Local Repeater

Let's program a local repeater near ARRL Headquarters. The W1AW repeater is managed by the Newington Amateur Radio League and offers coverage in the greater Hartford, CT area. The output frequency is 145.450 with a standard negative offset. The CTCSS tone is 127.3. Use memory Channel 004 for the entry.

1. Press **VFO/MR** to enter VFO mode. The radio will announce "Frequency Mode."
   - If the VFO indicator arrow appears beside the lower VFO, press **A/B** to select the upper display. VFO A is used to program the repeater channels into the radio. Programming data entered in VFO B (lower display) will not be saved.
2. Use the keypad to enter the desired frequency **1 4 5 4 5 0** for (e.g. 145.450 MHz). The radio will announce each number as it is entered.
3. Now let's add the minus 600 repeater shift.
   - Press **MENU** The radio will announce "Menu."
   - Navigate to menu option **26 OFFSET** using the ▲/▼ keys or by entering **2** then **6.**
   - Press **MENU.** The radio will announce "offset frequency."
   - Enter **0 0 0 6 0 0.** The radio will announce each number as it is entered.
   - Press **MENU** The radio will announce "confirm."
   - Press **EXIT.**

> ### Example Frequencies
>
> These repeater frequencies and offsets are just examples. If there are other repeaters you would like to program, feel free to substitute them. If you haven't already checked it out, Repeaterbook.com is a great place to go to learn about which repeaters are available in your area, what their frequencies, offsets, tones, and uses are. For more information on Repeaterbook, see the section 9 Importing Repeaterbook Entries to *CHIRP* in the next chapter.

5. Now we need to program the repeater offset direction.
    - Press **MENU.** The radio will announce "Menu."
    - Navigate to menu option 25 **SFT-D** using the ▲/▼ keys or by entering 2 then 5.
    - Press **MENU.** The radio will announce "frequency direction."
    - Select minus using the ▲/▼ keys.
    - Press **MENU.** The radio will announce "confirm."
    - Press **EXIT.**
6. Now we can set the transmit power level.
    - Press **MENU.** The radio will announce "Menu."
    - Navigate to menu option **2 TXP** using the ▲/▼ keys.
    - Press **MENU.** The radio will announce "power."
    - Select **HIGH** or **LOW** using the ▲/▼ keys.
    - Press **MENU** to confirm. The radio will announce "confirm."
    - Press **EXIT.**
7. Now we need to turn **TDR** to **OFF.**
    - Press **MENU.** The radio will announce "Menu."
    - Navigate to menu option 7 TDR using the up down keys or entering 7.
    - Press **MENU.** The radio will announce "dual standby."
    - Select **OFF** using the arrow keys.
    - Press **MENU.** The radio will announce "confirm."
    - Press **EXIT.**
8. Now let's program the CTCSS tone when we transmit.
    - Press **MENU.** The radio will announce "Menu."
    - Navigate to menu option 13 T-CTCSS.
    - Press **MENU.** The radio will announce "CTCSS."
    - Use arrow keys to select 127.3.
    - Press **MENU.** The radio will announce "confirm."
    - Press **EXIT.**

9. Now before we can enter the frequency into a memory channel we clear the target memory 004.
    - Press **MENU.** The radio will announce "Menu."
    - Navigate to menu option **28 DEL-CH.**
    - Press **MENU.** The radio will announce "Delete Channel."
    - Use arrow keys or enter **0 0 4.** The radio will announce each number as it is entered.
    - Press **MENU.** The radio will announce "confirm."
    - Press **EXIT.**
10. Now we can enter the VFO A frequency 145.450 into channel number 004.
    - Press **MENU.** The radio will announce "Menu."
    - Navigate to menu option **27 MEM-CH.**
    - Press **MENU.** The radio will announce "memory channel."
    - Enter **0 0 4** using the keypad or the ▲/▼ arrows. The radio will announce each number as it is entered.
    - Press **MENU.** The radio will announce "receiving memory."
    - Press **EXIT**

Now that you've programmed a repeater, try adding another on your own. Go ahead and add the W1OKY UHF repeater serving the Newington, CT area into memory channel 005 using these parameters: 443.050 MHz, + offset, with a tone of 100 Hz. If you need some help, you could start with step 3 above.

After programming the example frequencies, I would recommend programming all the local repeaters in your area and also adding any repeaters in places you intend to travel to in the near future. However, if there are more than a few repeaters you intend to program it's worth learning how to use the programming software *CHIRP*, which I cover in chapter 7. It makes the process much easier. That said, knowing how to program the radio manually is very useful when you don't have a computer handy.

Now, you're ready to attempt your first contact. Select a memory channel with an active local repeater and listen for a minute to make sure the repeater is clear. If you hear other hams talking, wait for an opening in the conversation before you attempt to transmit. If the repeater is clear, you can test it out. Hold your radio's microphone level with your mouth and be sure to keep the antenna pointed straight up. That will help direct your signal toward the repeater.

When you're ready, press and hold the PTT button. Speak slowly and clearly into the microphone something like, "Hello this is VE3IPS. I'm a new ham testing out my first handheld. Am I making it into the repeater?" It's important that you state your call sign, since transmitting without identifying is illegal (you must identify at least once every 10 minutes and at the end of the conversation). When you finish speaking, let go of the PTT button and listen for the repeater to acknowledge your transmission with a courtesy beep tone. If someone is monitoring the repeater they will come back to you and you can enjoy your first conversation with your new radio.

Congratulations on manually programming your first set of repeaters and (hopefully!) making your first contact.

6.6 Memory Navigation

Navigating memories on the Baofeng UV-5R is straightforward once you're familiar with the controls.

- Press the **VFO/MR** button to toggle between frequency mode and memory mode. The radio will announce "channel mode" or "frequency mode."
    - The display will show the frequency or channel name in the **VFO A** display.
- Stored channel numbers in memory mode appear to the right of the channel frequency.
- If you would like to listen on two different frequencies at once in channel mode, you can set **VFO B** to a saved channel and turn on **7 TDR**.

Scroll Through Channels

Use the ▲/▼ arrow keys on the keypad to browse through the memory channels sequentially. The radio will announce each channel number.

Direct Channel Entry

You can also enter the channel number directly on the keypad (e.g., pressing 1-2-3 will take you to channel 123), which is useful when your radio has many memory channels stored on it. The UV-5R can store 128 memory channels.

6.7 Settings Quick Start Guide

The radio SET menus can be a bit confusing for new users. I am including an overview of the common programmable features available on this radio with an explanation of their use.

MENU – 0 Squelch Settings

- Adjustable squelch levels (0 – 9) to filter (mute) out weak or unwanted signals. Set the squelch level from 0 (least squelch) to 9 (most selective). Lower settings allow weak signals to come through, while higher settings filter out noise. Typically, a setting of 2 – 4 is good for general use.

MENU – 1 Frequency Step

- Adjustable frequency step sizes for tuning:
  - 2.5, 5, 6.25, 10, 12.5, and 25 kHz.

MENU – 2 Transmit Power Levels

- High Power: 4 – 5 watts
- Low Power: ~1 watt
- Can be set per channel or in VFO mode.

MENU – 3 Battery Saver Mode

- Automatically reduces power consumption when no signals are detected.

MENU – 4 VOX (Voice-Activated Transmission)

- Enables hands-free operation by transmitting when it detects your voice.
- Adjustable VOX sensitivity levels.

MENU – 7 Dual Watch/Dual Standby

- Monitors two frequencies or channels simultaneously.
- Only one can transmit, but both are monitored for activity.

MENU – 8 Roger Beep

- The UV-5R can beep after every transmission (roger beep), which may not be necessary in most situations but can be useful for simplex transmissions.

MENU – 9 Time-Out Timer (TOT)

- Limits the maximum duration of continuous transmission (default is 60 seconds).
- Prevents overheating and accidental long transmissions.

MENU – 11 and 13 CTCSS and DCS Tones

- CTCSS (Continuous Tone-Coded Squelch System)
  - 50 standard tones for repeater access or pseudo-private conversations.
- DCS (Digital-Coded Squelch)
  - 105 standard codes for advanced squelch filtering.

MENU – 18 SCAN Properties

- If you're searching for activity on the bands, you can:
  - **Scan Channels:** You can scan programmed channels by pressing and holding the */SCAN button.
  - **VFO Scan:** In VFO mode, the UV-5R will scan through a frequency range.
  - **Stop Scanning:** The radio will stop scanning when it detects activity. To stop the scan manually, press any key.
  - The **CO** setting will allow the SCAN to resume once the transmission has stopped. You can also set it to resume scanning after a few seconds of a found signal (**TO**).

MENU – 21 Display Mode VFO A

- You can choose what the display will show: Frequency, Channel Number or Name.
  - I use NAME for this VFO.

MENU – 22 Display Mode VFO B

- You can choose what the display will show: Frequency, Channel Number or Name.
  - I use FREQ for this VFO.

MENU – 23 Busy Channel Lockout (BCL)

- Prevents transmission on a busy frequency or channel.

MENU – 24 Keypad Lock

- Programmable lock to prevent accidental keypresses.

MENU – 25 and 26 Repeater Settings

- **Offset (Repeater Shift):**
  - Allows setting of the offset frequency for accessing repeaters (up to 69.995 MHz).
- **Offset Direction:**
  - Configure + or - shifts for repeater use.

MENU – 32 Alarm Function

- The UV-5R has an emergency alarm feature that can send out a loud tone or transmit a signal. This can be useful in real emergencies, but use it with caution as it transmits on whatever frequency you are tuned to and does not identify, so it can be considered QRM (manmade noise).
- To activate the emergency alarm, hold the orange **CALL** button for a few seconds.

- Includes three modes:
  - **Site Alarm:** Sounds an alarm locally.
  - **Remote Alarm:** Transmits an alarm tone to other radios.
  - **SOS Mode:** Alerts nearby operators.

MENU – 40 Reset

- If something isn't working properly or you want to start with a clean radio, you can perform a reset. Be careful, as this will erase all programmed channels and settings, returning the radio to its default state.

6.8 UV-5R Programming and Usability Tips

The UV-5R is a feature rich radio. The tips below can help you maximize the usability and functionality of your radio.

6.8.1 Use the Dual VFO Effectively

The UV-5R has a dual VFO feature, allowing you to monitor two frequencies at once. To switch between the two:

- Press the **A/B** button to switch between the VFO A (upper) and VFO B (lower) frequency.
- The **TDR** (dual watch mode) allows you to monitor both frequencies and automatically receive a transmission on either one. You can turn it on in menu **7 TDR**.

6.8.2 Understand Power Settings

The UV-5R has two power levels low (1W) and high (4 – 5W). You can switch between them in Menu **2 TXP.**

- Use low power when you're close to a repeater or communicating via simplex over short distances to conserve battery.
- Use high power when you're farther away from a repeater or need to maximize your range.

6.8.3 Battery Saving Tips

- **Battery Saver:** Enable the battery saver in **Menu 3 (SAVE).** This will put the radio into a low-power standby mode when it's not actively transmitting or receiving.
- **Keypad Lock:** To prevent accidentally pressing buttons and draining your battery, lock the keypad by holding the # key for a few seconds. Repeat to unlock.

6.8.4 Use the FM Radio Feature

The UV-5R has an FM radio mode that can be handy when you're not using it for communication.

- Press the orange **CALL** button to switch to FM broadcast radio (88 – 108 MHz).
- You can still monitor your primary frequency while using the FM radio. If a signal is received, the UV-5R will mute the FM radio automatically.
- Program FM broadcast stations into channels for easy access.

6.8.5 Improve Audio and Transmission Quality

- **External Speaker/Mic:** Consider using an external speaker and mic for clearer audio when transmitting or receiving in noisy environments.
- **Monitor Transmit Audio:** Ask for a signal report to make sure your transmitted audio sounds clear and strong. If your voice sounds muffled or low, consider speaking more directly into the mic or holding it closer to your lips.

6.8.6 Tone Scanning

The **Tone Scanning** feature that allows users to discover the CTCSS or DCS tone being used on a particular frequency while the station is transmitting. This is especially useful when trying to identify the tone required to access a repeater (it may not be listed publicly or it has been changed) or to match the tone of a received transmission.

1. Tune to the desired frequency.
   - Switch to **Frequency Mode** (VFO mode) or select a programmed channel with the desired frequency.
2. Activate tone scanning.
   - Press the **MENU** button.
   - Navigate to the **MENU 11 (T-CTCS)** or **MENU 13 (T-DCS)** option using the up/down arrow keys. These correspond to the tone squelch settings.
   - Press **MENU** again to enter the setting, and then press the *[SCAN]** button to begin scanning for the tone.
3. Observe the display.:
   - The radio will scan through all possible **CTCSS** or **DCS** tones to find the matching one.
   - Once a matching tone is found, it will display on the screen.
4. Save the tone.
   - If you want to store the identified tone, press **MENU** to confirm, then save the configuration.
5. Exit the menu.
   - Press **EXIT** to return to normal operation.

6.8.7 LED Flashlight

The built-in flashlight with programmable options for steady or strobe modes is actually quite useful.

6.9 Usability Review

The Baofeng UV-5R is designed to meet the needs of both beginners and experienced users. Although the manual can be somewhat confusing, this guidebook aims to simplify the learning process and help you get the most out of your radio.

The radio itself feels solid, with a comfortable physical layout. However, the buttons might feel small for users with larger hands. The dual-band display is clear and easy to read, and all buttons offer a satisfying tactile response. Voice prompts make navigating the menus straightforward, though this feature can be turned off if preferred.

In terms of performance, the Baofeng UV-5R performs well in most environments. Audio clarity is good for both transmission and reception, although it may not match the quality of more expensive models. The range is comparable to other radios in its class, with VHF frequencies typically performing better over longer distances than UHF.

I've dropped these radios several times, and while they've picked up a few scuffs and scratches, they've continued to operate without issue. The belt clip, which I initially doubted, has proven to be both sturdy and reliable.

While the UV-5R is durable, it's not the most weather-resistant option available. For use in extreme weather—such as heavy rain or snow—you might want to consider investing in a protective cover or opting for a more weatherproof model, such as one with an ingress protection (IP) rating.

The radio is powered by a rechargeable 1800 mAh Li-ion battery, which provides several hours of continuous operation on a single charge. Depending on usage, the battery can last several days on standby or a full day with regular use. The included desktop charging dock makes recharging simple and convenient.

Users have reported success in various scenarios, including local communications, simplex operations, and repeaters, making it a reliable option for general amateur radio use. However, the radio is prone to interference, particularly in densely populated urban areas. This is due to its broad RF front end, which allows a wide range of frequencies for transmission. By contrast, radios specifically designed for amateur bands often feature better front-end filtering to reduce interference. For those operating in less RF-dense environments, the UV-5R performs as expected without major issues.

In my years of using Baofeng radios, I've found that the performance-to-cost ratio is exceptionally high. However, affordability is meaningless if the radio doesn't perform well. Baofeng continues to deliver high-performance radios at attractive prices, with the UV-5R remaining a popular and reliable choice.

6.10 Accessories

The Baofeng UV-5R is a versatile and popular handheld transceiver, and several accessories can enhance its functionality and usability.

6.10.1 Extra Batteries

- **Higher Capacity Batteries:** Consider purchasing high-capacity batteries (BL –5L 3800 mAh) for extended operating time, especially during outdoor activities.

6.10.2 Charger Options

- **Desktop Charger:** A dual or multi-charger can charge multiple batteries or radios simultaneously.
- **USB Charging Cable:** Allows you to charge your UV-5R from a power bank or USB port, useful for portable operations.

6.10.3 Antenna Upgrades

- **Better Antennas:** Upgrading to a longer antenna (like a Nagoya NA-771 or Diamond SRH77CA) can improve reception and transmission range.
- **SMA-Female to SMA-Male Adapters:** These can help connect various antennas to the UV-5R.

6.10.4 Headsets and Microphones

- **Earpiece with PTT:** A simple earpiece with a push-to-talk (PTT) button allows for discreet communication and is ideal for noisy environments.
- **Boom Microphone Headset:** Provides clearer audio for both transmitting and receiving, particularly in high-noise areas.

6.10.5 Carrying Cases and Straps

- **Protective Cases:** A soft or hard case can protect your UV-5R from scratches and damage.
- **Shoulder Straps:** A comfortable strap or lanyard allows for easy carrying during extended use.

6.10.6 USB Programming Cables

- Essential for programming frequencies, settings, and other configurations using software like *CHIRP*. Baofeng makes the K-Plug Programming Cable. Competitors like the RT Systems, USB-K4Y cable also work well.

6.10.7 Software for Programming

- ***CHIRP* Software:** Free software that allows you to easily manage frequencies and settings on your UV-5R through a computer.
- **RT Systems:** BAO-5R programming software with K4Y-USB programming cable works well.

6.10.8 External Speakers

- A small external speaker can improve audio quality and make it easier to hear incoming transmissions.

6.10.9 Mobile Mounts and Holders

- If using the UV-5R in a vehicle, a dash or cup holder mount can keep it secure and accessible. Lido is a respected manufacturer of all kinds of vehicle mounts.

6.10.10 Battery Eliminators

- Allows you to power the UV-5R from your vehicle's power supply, eliminating the need for batteries during long trips.

These accessories can significantly enhance the functionality and convenience of the Baofeng UV-5R, whether you're using it for casual communication, emergency preparedness, or outdoor activities.

6.11 Cheat Sheet

Below, I have compiled a quick use cheat sheet. It includes information on the available options for each menu item and provides the factory defaults as well as commentary on how I use the radio.

Menu #	Menu Item	Function	Options/Default
0	SQL	Adjust squelch level to filter weak signals by muting the recover	0 (Off) – 9 (Tight); Default: 5
1	STEP	Set frequency step size for tuning. 5 kHz for VHF and 25 kHz for UHF	2.5/5/6.25/10/12.5/25 kHz; Default: 2.5
2	TXP	Set transmission power level for 1 or 4 watts.	HIGH/LOW; Default: HIGH
3	SAVE	Enable/disable battery saver mode to extend battery life.	OFF/1/2/3/4; Default: 3
4	VOX	Enable hands free voice-activated transmission and adjust sensitivity level.	0 (Off) - 10 (High); Default: 0
5	WN	Set narrowband (NARROW) or wideband (WIDE) mode for channel bandwidth.	WIDE/NARR; Default: WIDE
6	ABR	Set time duration for the display backlight to remain on.	1 – 10 sec/ON/OFF; Default: 5 sec
7	TDR	Enable/disable dual-watch to monitor two frequencies simultaneously.	ON/OFF; Default: OFF
8	BEEP	Enable/disable keypad tone when pressing buttons.	ON/OFF; Default: ON
9	TOT	Set transmission time-out timer to limit continuous transmission. I set mine to 3 minutes.	15 – 600 sec; Default: 60 sec

Menu #	Menu Item	Function	Options/Default
10	R-DCS	Set receive DCS (digital-coded squelch).	DCS codes; Default: OFF
11	R-CTCS	Set receive CTCSS (continuous tone-coded squelch system).	CTCSS tones; Default: OFF
12	T-DCS	Set transmit DCS tone for repeater access or private communication.	DCS codes; Default: OFF
13	T-CTCS	Set transmit CTCSS tone for repeater access or private communication.	CTCSS tones; Default: OFF
14	VOICE	Select voice prompt language when pressing buttons.	OFF/CHINESE/ENGLISH; Default: ENGLISH
15	ANI-ID	Program the Automatic Number Identification (ANI) code for your radio. Not used by hams.	1-9999999; Default: Blank
16	DTMFST	Enable/disable DTMF tone during transmission.	OFF/DT+ANI/ANI+DT/DTMF; Default: DT+ANI
17	S-CODE	Set the stored DTMF signaling codes for call and control operations.	1-15; Default: 1
18	SC-REV	Set scan mode (resume after activity, pause, or carrier). I prefer CO as the repeater will scan at the end of the transmission.	TO/CO/SE; Default: TO
19	PTT-ID	Configure when the radio sends the ANI-ID. Not Used by hams.	OFF/BOT/EOT/BOTH; Default: OFF
20	PTT-LT	Set the delay time for the PTT ANI-ID. Not used by hams.	0-30 ms; Default: 5 ms
21	MDF-A	Set display mode for Channel A (frequency, channel number, or name). I prefer NAME.	FREQ/CH/NAME; Default: FREQ
22	MDF-B	Set display mode for Channel B (frequency, channel number, or name). I prefer FREQ.	FREQ/CH/NAME; Default: FREQ
23	BCL	Enable busy channel lockout to prevent transmission on an active channel.	ON/OFF; Default: OFF
24	AUTOLK	Automatically lock keypad after a set period of inactivity.	ON/OFF; Default: OFF
25	SFT-D	Set repeater offset (splits) direction. Use OFF for simplex.	OFF/+/−; Default: OFF
26	OFFSET	Set repeater offset frequency for duplex operation. This is uses a 600 KHz (VHF) and 5 MHz (UHF) standard.	0-69.995 MHz; Default: 00.000 MHz
27	MEM-CH	Save current settings to a memory channel.	0 – 127
28	DEL-CH	Delete a programmed memory channel.	0 – 127
29	WT-LED	Set color of display backlight when receiving a transmission.	Purple/Orange/Blue; Default: Purple

Menu #	Menu Item	Function	Options/Default
30	RX-LED	Set color of display backlight when receiving a signal.	Purple/Orange/Blue; Default: Blue
31	TX-LED	Set color of display backlight when transmitting.	Purple/Orange/Blue; Default: Orange
32	AL-MOD	Set the alarm mode (local, remote, or SOS).	SITE/TONE/CODE; Default: SITE
33	BAND	Set frequency band for the VFO.	VHF/UHF
34	TDR-AB	Set which channel (A or B) is used for transmission during dual-watch.	A/B; Default: A
35	STE	Enable/disable squelch tail elimination (STE) during transmission.	ON/OFF; Default: ON
36	RP-STE	Set the squelch tail elimination time.	1 – 10 (100 ms steps); Default: 5
37	RPT-RL	Enable/disable squelch tail elimination for repeaters.	ON/OFF; Default: OFF
38	PONMSG	Set power-on message display (custom text or voltage). I use BATTERY.	MSG/BATTERY
39	ROGER	Enable/disable the Roger beep tone after transmission. This is useful for Simplex operations but do not use on repeaters.	ON/OFF; Default: OFF
40	RESET	Perform a factory reset of the radio.	VFO/ALL

7 Programming the UV-82 with *CHIRP*

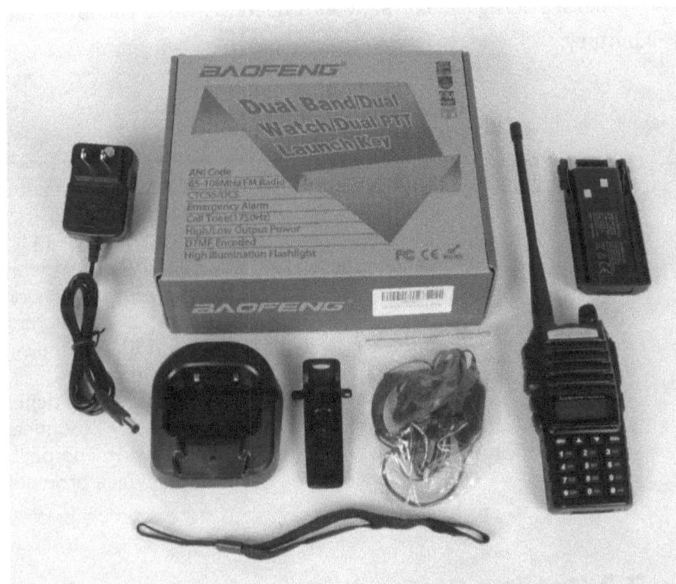

7.1 Unboxing

The UV-82 improves upon the UV-5R with a louder speaker, a more solid case, larger buttons, and more output power. It is not simply a remake but a new radio from the ground up.

The accessories such as the batteries are radio specific, but other accessories like external antennas, speakers, and mics are common across most Baofeng models.

The contents of the Baofeng UV-82 box vary depending on the package, but typically include the following items:

- **Baofeng UV-82 Radio** — The handheld transceiver itself.
- **Antenna** — A flexible 5" dual-band antenna (VHF/UHF) with an SMA-Female connector.
- **Battery Pack** — A rechargeable lithium-ion (Li-ion) battery, typically rated at 2,000 mAh.
- **Belt Clip** — A screw-on clip for attaching the radio to your belt or pocket.
- **Hand Strap** — A lanyard-style wrist strap for additional portability.
- **Desktop Charger** — A cradle-style drop-in charger with an ac adapter for charging the battery.
- **Earphone/Microphone** — A single, open, over-the-ear style headphone with dual PTT buttons.
- **User Manual** — A basic instruction manual that explains the radio's features, programming, and operation.

Programming Cable

If you don't already have one, consider purchasing a Baofeng K-Plug programming cable. Without one, or a similar programming cable, you can't use software to program your radio, which is one of the main thing's I'll cover in this chapter.

Before using the UV-82, you'll need to attach the antenna and battery pack and charge the battery.

Table 7.1
Basic Specification

Frequency Range	65 – 108 MHz (Commercial FM) 136 – 174 MHz (VHF) 400 – 520 MHz (UHF)
Power Output	1W / 5W / 7W or 8W selectable
Channels	128 programmable channels
Battery	2000 mAh Li-ion rechargeable battery (7.4V)
Antenna	Dual Band SMA Female
Display	Dual band display
Dimensions	4.33 inches x 2.36 inches x 1.26 inches.
Weight	Approximately 7.5 ounces (including battery)
Connectivity	2-pin Kenwood-compatible audio jack
Additional Features	Flashlight, voice prompts, and scan functions

7.2 Getting Started

The Baofeng UV-82 is a versatile handheld and is basically a UV-5R with some worthwhile feature improvements.

The following are the differences between the UV-82 and the UV-5R:

- The UV-82 is about half an inch taller than the UV-5R.
- The UV-82 has two separate LEDs located on the top panel.
- The **VFO/MR** key on the UV-5R has been replaced by holding **MENU** while turning on the radio. This switches the radio from VFO mode to memory mode and vice versa.
- The **Band** key on the UV-5R is now selected in the menu.
- The UV-82 has dual PTT keys. Press the upper one to transmit on VFO A, press the lower one to transmit on VFO B.
- The UV-82 has a slightly higher advertised power output (8 W on 2 meters and 7 W on 70 cm).
- You can use your UV-5R antenna and speaker/mic accessories but not the battery pack with the UV-82.

After unboxing, you will need to charge the battery. The charge time is approximately 5 hours for a full charge for the 2,000 mAh BL-8 battery.

The battery charge indicator will glow solid red while charging and solid green when the battery is fully charged. If the charge indicator flashes red and green then it may indicate a faulty battery pack. Please be aware that a Li-ion battery will be at full strength only after you have depleted and charged it least three times. After these three cycles you can charge the battery any time you want. There's no memory effect to deal with. However, a word of caution: **Do not leave the radio in the charger for long periods of time when the battery is charged.** Li-ion batteries can overcharge and explode. Most chargers, batteries, or both have built in charge control circuits that should prevent overcharging, but Li-ion fires are extremely dangerous because the battery chemistry produces its own fuel for the fire. It's best to take reasonable precautions to avoid catastrophic failure.

After you've charged the battery, install the belt clip on the radio using the supplied screws. There are two, one on either side of back of the radio. Next, remove the antenna from its wrapping and twist it into place on the SMA jack opposite the volume dial on top of the radio. (For a list of the radio's components, consult the numbered photograph of the radio below.) Do not overtighten the antenna or you could damage it or the transceiver. Afte the antenna is in place, slide the battery pack under the clip and give it a little force until it clicks into place. Finally, twist the volume dial until it clicks, usually less than an eighth turn. This turns the radio on. Using the side F button, you can listen to a local FM broadcast station while I walk you through the steps to program the radio. Pressing the up or down arrow adjusts the radio's receive frequency.

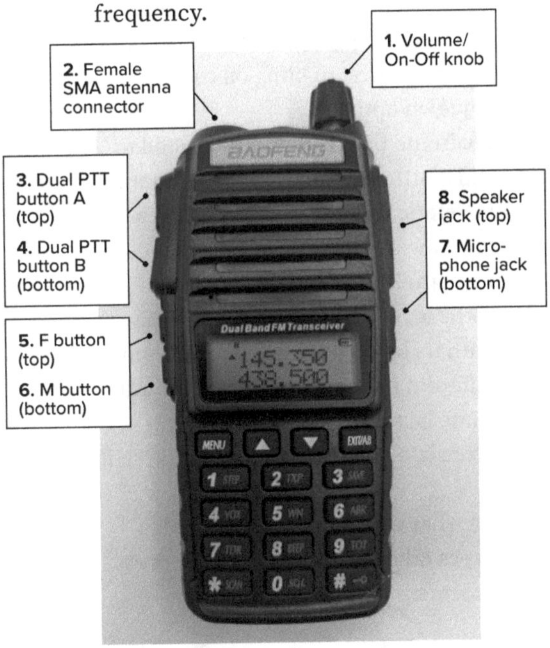

7.3 Button by Button Walkthrough

7.3.1 The Front of the Radio

FRONT PANEL

The UV-82 has an informative two-line LCD display that shows key information such as the frequency, battery level, signal strength, mode indicators, power level, and any activated features (like CTCSS tones). The display backlight can be set to different colors for various modes.

KEYPAD

The keypad is a 16-button grid that includes numbers and several multifunction keys. Each key serves various purposes for navigation, menu settings, and direct frequency input. The **MENU**, ▲ / ▼ Arrow keys, and **EXIT/AB** serve additional functions.

The **0 – 9** number keys are multi-purpose. When in the Frequency mode, which the radio defaults to when turned on, pressing the number buttons programs whichever frequency line is selected on the radio. An arrow appearing to the left of the numeric frequency indicates that frequency line is selected. To change frequency lines, press the **EXIT/AB** button. The arrow indicator should shift from one frequency line to the other.

When navigating the radio's menu, the keypad keys act as shortcuts to 11 often-used functions out of the 41 numbered functions. To access the radio's menu simply press the **MENU** button. This will bring you to the last menu item you were using. When first turning on the radio, it will bring up the first menu item, which is **0 / SQL** or the squelch option.

Before I move on to discussing the menu options available in the UV-82, it might be good to pause on the squelch feature, since it's useful, but might need a little explanation for someone new to radio. Increasing the squelch decreases the amount of static that the receiver admits. This can be useful when in an RF noisy area, but you should be aware that the higher you set the squelch, the more powerful a signal must be to break through. If your squelch is set too high, you run the risk of missing signals that may be nearer the level of the noise. To figure out what squelch level is appropriate, you need to experiment with its levels during use. You should also know that the appropriate level may change and that different use cases require different squelch levels. Just because a squelch level of 3 was appropriate last time you accessed a repeater doesn't mean that the same squelch level will be appropriate if you're hunting weak VHF or UHF signals or trying to make contacts on a satellite.

All right, enough about squelch. Play around with it to see what works for you. Onward to the UV-82's menu features. Here are the shortcuts the **0 – 9** keys take you to with a single press (when in the menu) and the actions the row of four keys performs:

0 / SQL	Access the squelch level settings. Levels are from 0 to 9, with 9 being the highest level of squelch. - To view the battery's voltage level, hold **0** at any time while the radio is on.
1 / STEP	Set frequency step size for tuning. This is the number of Hz or kHz each button click moves the frequency up or down. Bigger steps are generally better for higher frequencies. A step of 12.5 kHz is reasonable for most VHF frequencies.
2 / TXP	Toggle transmit power between Low (1 W), Medium (5 W), and High (7 or 8 W) in the menu.
3 / SAVE	Enable/disable battery saver mode to extend battery life. - To view the firmware version, hold 3 while powering on the radio.
4 / VOX	Enable hands-free voice-activated transmission and adjust sensitivity level.
5 / WN	Set narrowband (NARROW) or wideband (WIDE) mode for channel bandwidth. - Amateur radio generally uses wide FM (25 kHz bandwidth), especially on repeaters and for simplex, while GMRS and FRS radios use narrow FM (12.5 kHz), which allows for more channels at the expense of sound quality.
6 / ABR	Set the length of time the display backlight remains on. - To view the manufacture or firmware date (there's no consensus on what this date actually means but it's likely a manufacture date), hold 6 while powering on the radio.
7 / TDR	Turn dual watch/dual reception on or off. Turning off **TDR** prevents reception on the secondary channel, though it does not prevent transmission.
8 / BEEP	Enable/disable keypad tone when pressing buttons. - To view the radio's model number, hold 8 while powering on the radio.
9 / TOT	Set transmission time-out timer to limit continuous transmission.
*** / SCA**	Long press to initiate frequency or channel scanning. In VFO mode, the radio will scan through a frequency range. In MR mode (memory recall), the radio will scan through the programmed memory channels. To stop the scan, press any key, including the PTT buttons.
# / LOCK	Long press to lock/unlock the keypad. (This function is not a menu option and works in both Frequency and Menu modes.)
MENU	Opens the radio's settings menu. Used to confirm selections in programming mode.
▲ / ▼ (Arrow Keys)	Scroll through menu options or frequencies. Navigate stored memory channels in channel mode.
EXIT/AB	Exits the menu or cancels an action. Also switches between the upper and lower frequency line.

7.3.2 Special Buttons on the Side

PTT BUTTONS (DUAL PTT)

PTT A Transmit on the upper display frequency/channel.

PTT B Transmit on the lower display frequency/channel.

F: FM RECEIVER AND ALARM

The UV-82 comes with broadcast FM receive capability. The receive range is from 65 – 108 MHz. The [F] button shifts the receiver into the FM broadcast frequency range. It also provides access to the alarm function.

- A short press turns the FM receiver on or off. When turned off, the receiver returns to the last used frequency on the 2-meter band and/or the 70-centimeter band.
- A long press activates and deactivates the **ALARM** function. Unlike the UV-5R discussed earlier, the UV-82 provides three options for the Alarm:
  - **Site** – This mode does not transmit, but makes an alarm play through the radio's speaker.
  - **Tone** – This mode transmits a cycling tone over the air. I would not recommend using this feature, likely even in a real emergency, as it provides no information other than an **ANI ID** if you have programmed one in, and it requires radio direction finding to figure out where you are. If it isn't a real emergency, then this mode only generates manmade RF noise (called QRM), which is illegal.
  - **Code** – This mode transmits a code 1-1-9 followed by the ANI code. Again, I would recommend setting the alarm to Site mode, so that you don't run afoul of FCC regulations.

M: MONITOR

- This button toggles the flashlight between on (one press), flashing (two presses), and off (three presses).
- A long press deactivates the squelch, allowing you to monitor a frequency for signals down near the noise floor.

7.3.3 Top Controls and Indicators

ANTENNA CONNECTOR

The UV-82 has a removable antenna connected to a standard SMA female port. This allows users to swap the stock antenna with compatible aftermarket antennas for better performance as long as they are antennas with SMA male connectors.

VOLUME KNOB (POWER SWITCH)

The top rotary knob controls the radio's volume and serves as the power switch.

LED INDICATOR

Two small LEDs at the top glow different colors to indicate the radio's status:

- Right LED: Red indicates transmitting.
- Left LED: Green indicates receiving a signal.

FLASHLIGHT

The UV-82 includes a small LED flashlight, activated by a quick press on the **MONI** button (side button **M**). It's not very bright but can be helpful in low-light conditions. For directions, see section 7.3.2.

7.3.4 Ports

SPEAKER/MIC PORT

Located on the side of the radio, this dual-pin port allows you to connect external accessories such as a speaker, mic, or programming cable. It uses the Kenwood radio standard with a 2.5 mm top pin for the speaker and a 3.5 mm bottom pin for the microphone. Use this side port and a compatible cable to connect the radio to a computer for programming with software like *CHIRP*. *CHIRP*, or similar programming software, is the easiest way to manage channels, frequencies, and other settings. Due to the UV-82's dual PTT design, if you choose to use a headset, it will need to have two PTT buttons.

7.3.5 Battery Pack and Belt Clip

The UV-82 uses a removable Li-ion battery pack that locks into the back of the radio. This design allows for easy replacement and upgrading to higher-capacity battery packs. The sturdy belt clip attaches to the radio with two small cross screws.

7.4 Programming

7.4.1 Manual Programming

In the previous chapter, I covered programming the UV-5R manually, and the process for the UV-82 is similar, with a few key differences noted below. If you own a UV-82 and are reading this chapter without having read the chapter on the UV-5R, read through the section below on manual programming to note the differences between the two radios, then go back to chapter 6 for detailed walkthroughs for manually programming many of your radio's features.

The major difference from the UV-5R is that all programming must be done in the **Frequency Mode** using the **Upper Display** only. From there you may assign the entered data to a specific channel for later access in the **Channel Mode** if desired.

Programming done using the **Lower Display** cannot be saved as a **Channel Mode** and will be lost.

To illustrate this, I'll walk you through programming a simplex weather channel frequency on the Baofeng UV-82.

1. Turn the **Volume/On-Off** knob to power on the radio..
   - To tell if you are in Channel Mode, look for a channel number to the right of either the **Upper or Lower Display** frequency. If you see a number beside the frequency, you are in channel mode. You cannot input frequencies in **Channel Mode.**
   - To exit **Channel Mode** turn off the radio. Hold down the **MENU** key while you twist the **Volume/On-Off** knob. Continue holding the **MENU** key until the radio is fully powered on. You should see a pair of frequencies in the display. To their right only the battery level indicator should appear. If you still see numbers beside the displayed frequencies, you may not have held down the **MENU** key through the whole power-on cycle. Turn off the radio and try again, making sure to hold down the **MENU** key until frequencies display.
   - With **Voice Menu** enabled, the radio will say aloud each button entry after you push it.
2. Set the frequency 1 6 2 4 0 0 using the number keys. This sets the frequency to the number you entered.

You're now ready to use your Baofeng UV-82 for listening to the NWS weather in your area. Programming other simplex frequencies uses the same process, and with the adjustments noted, programming repeater uplink, downlink, and tone frequencies is the same as the process we went through earlier. See chapter 6 for more detail. For the rest of the programming information, I will use *CHIRP.*

7.4.2 Software Programming Options

Both radios support 128 memory channels. Programming all of them is a big task if you're going to try to do it manually. Using software speeds up the process, increases the entries' accuracy, and allows you to back up your radio on your computer. I should also mention that the process for programming the UV-5R is identical to the one for programming the UV-82, so even if you don't have the UV-82, you'll want to work through the steps in this chapter to learn about programming your radio's memory using *CHIRP.*

CHIRP SOFTWARE

This software makes programming Baofeng radios—and other brands—simple and efficient. It supports naming memory channels, importing/exporting data, and integrating with resources like Repeater Book. You can also program a variety of radio settings with this software and prepare multiple files for uploading or to save as back ups. It is updated weekly with bug fixes and support for new radios.

I like to maintain separate files for different deployments. For instance, I have a travel radio set up for Connecticut, Massachusetts, Vermont, New Hampshire, and New York. Additionally, I have a California file that covers the greater San Francisco area (including San Jose and Napa Valley) and a separate Los Angeles file. My Canadian file includes repeaters from coast to coast.

Before traveling, I can easily upload a specific file and switch back to my standard local memory file when I return. To streamline my travel setup, I've dedicated a UV-5R as a travel radio, filling its 128 memory channels with my favorite frequencies across the United States.

Baofeng Customer Programming Software

Baofeng also provides its own proprietary programming software that supports specific Baofeng radios. It basically does what *CHIRP* does but is more limited in its feature set. I use it as well as *CHIRP*. The organization of the settings screen makes it easy to see an overview of the features that you can adjust.

Download the software from the official Baofeng website at **www.baofengradio.com**.

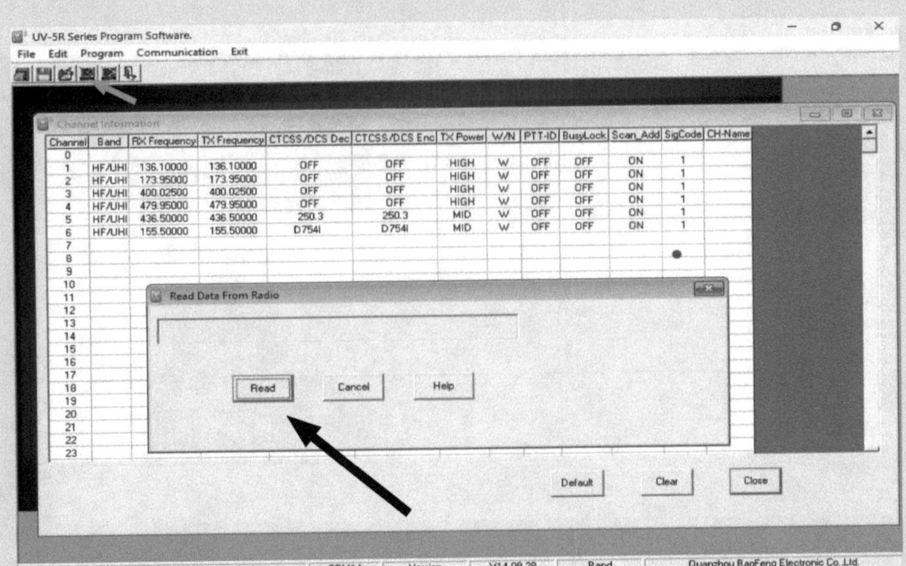

continued

Baofeng Customer Programming Software

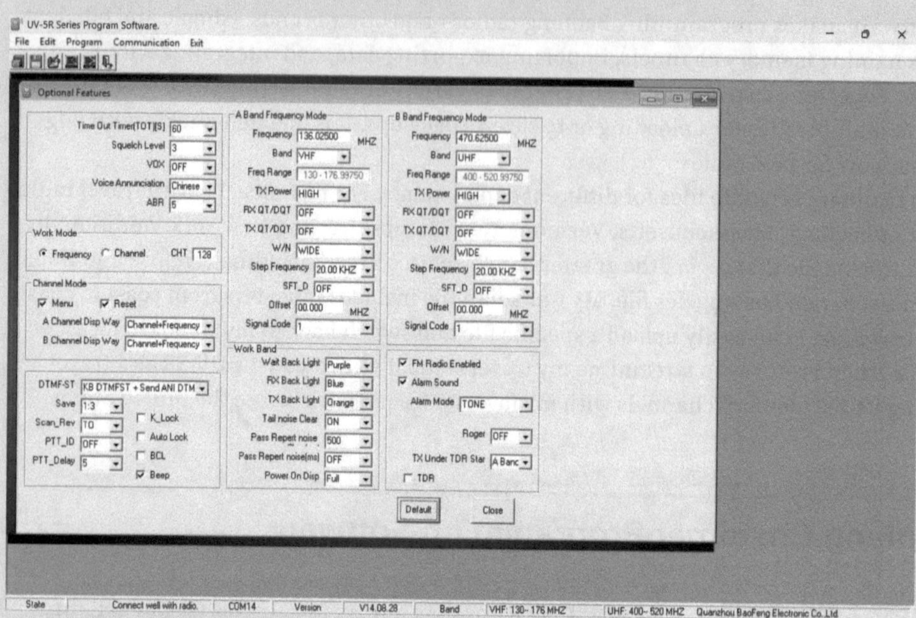

I like the way the settings are displayed. It makes selection easy and at times I will use this software for programming instead of *CHIRP*.

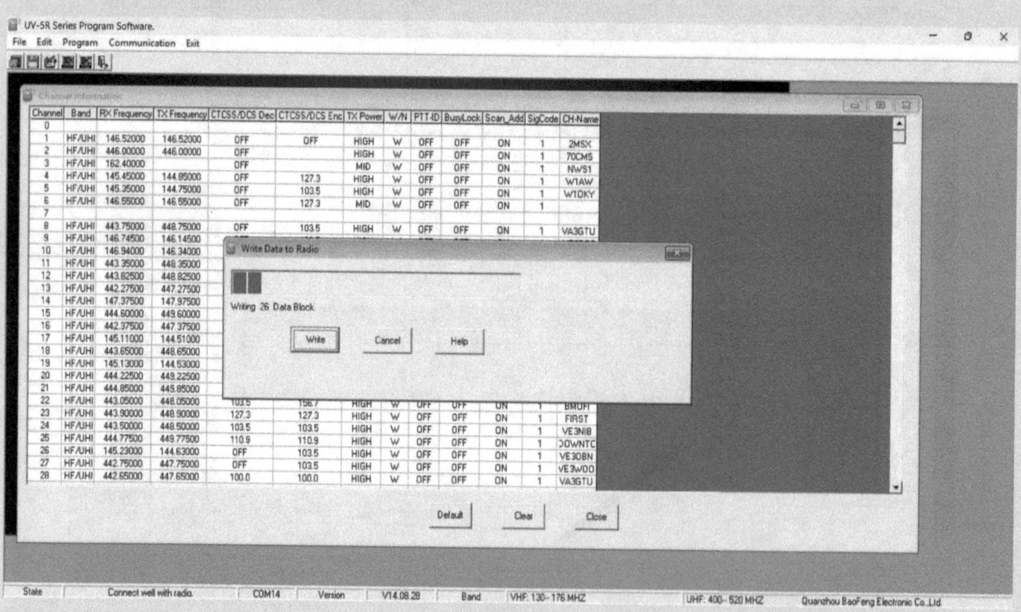

7.4.3 Interface Programming Cables

I strongly recommend purchasing a programming cable with your radio from an authorized seller or direct from **www.Baofengradios.com.**

When I first got my radio, I ended up getting the programming cable from another source because it was cheaper by a few dollars. I learned that buying cheap meant buying twice after a few hours of frustration trying to get a bad cable to work.

To help you avoid cable problems, here are a couple troubleshooting tips:

- **Programming Cable:** Ensure you are using the correct manufacturer cable. Many third party cables don't work.
- **Driver Issues:** If the radio isn't recognized, ensure you've installed the correct USB drivers and that the radio is powered on.
- **COM Port Conflicts:** Check your *Device Manager* (Windows) to confirm the COM port assigned to your radio and make sure it's selected in the *CHIRP* radio settings tab.

I have also been using the RT Systems K4Y-USB programming cable as it supports a variety of brands and radios. It has been proven to work on Windows 11. The drivers are available on the RT Systems website (**www.rtsystemsinc.com**) along with installation support.

7.4.4 Programming the UV-82 or the UV-5R Using *CHIRP*

Programming the UV-82 manually uses the same procedure as the UV-5R. So, instead of detailing that here, I'll focus on using the computer program *CHIRP* to program the UV-82. If you're looking for directions on manually programming the UV-82, check out the directions in chapter 6. I should also note that you can program the UV-5R with *CHIRP* too, and the procedure is identical except for character limitations and the like. When those differences crop up, I'll let you know.

CHIRP is a free, open-source software designed to simplify the programming of Baofeng radios and many other brand models. It offers a user-friendly, spreadsheet-like interface for managing frequencies, tones, offsets, power levels, and channel names, making it far easier than manual radio programming. It's compatible with Windows, macOS, and Linux and is regularly updated to support new radios. With *CHIRP*, users can download their radio's current configuration, edit it, and upload new settings in bulk. *CHIRP* also allows channel management, making it easy to organize channels by importing, exporting, and copying frequency lists in CSV. This is useful when programming multiple channels or importing lists from repeater listing resources. Additionally, *CHIRP* allows you to back up your radio's configuration, ensuring settings can be restored if needed. To program your radio with *CHIRP*, you'll also need a programming cable (preferably one with reliable drivers, FTDI or Silicon Labs).

CHIRP Software Development

Development of *CHIRP* is an all-volunteer effort and is offered as open-source software, free of charge. If you like *CHIRP*, please consider contributing a small donation to help keep it going.

Before proceeding, if you are part of an amateur radio club, it might be worthwhile to ask its members if they have a shared *CHIRP* file with the most used local frequencies already loaded. If they do, then you can modify or add to the file to meet your specific needs.

CHIRP is a valuable tool for amateur radio enthusiasts, offering efficiency, flexibility, and precision for programming a variety of radios.

CHIRP SOFTWARE INSTALLATION

The following procedure walks you through installing and configuring *CHIRP* on your PC and gets you started with programming your radio. If you encounter any issues, the CHIRP FAQ (**chirpmyradio.com/projects/chirp/wiki/How_To_Get_Help**) and forums can provide additional help.

In this chapter we will install and use *CHIRP* on a Windows computer. The steps for other operating systems are similar but have different installation syntax commands.

STEP 1: DOWNLOAD *CHIRP*

1. Go to the official *CHIRP* website at **chirpmyradio.com/projects/chirp/wiki/Download**
2. Click on the **Download** link in the menu.
3. Select the version of *CHIRP* compatible with your operating system.

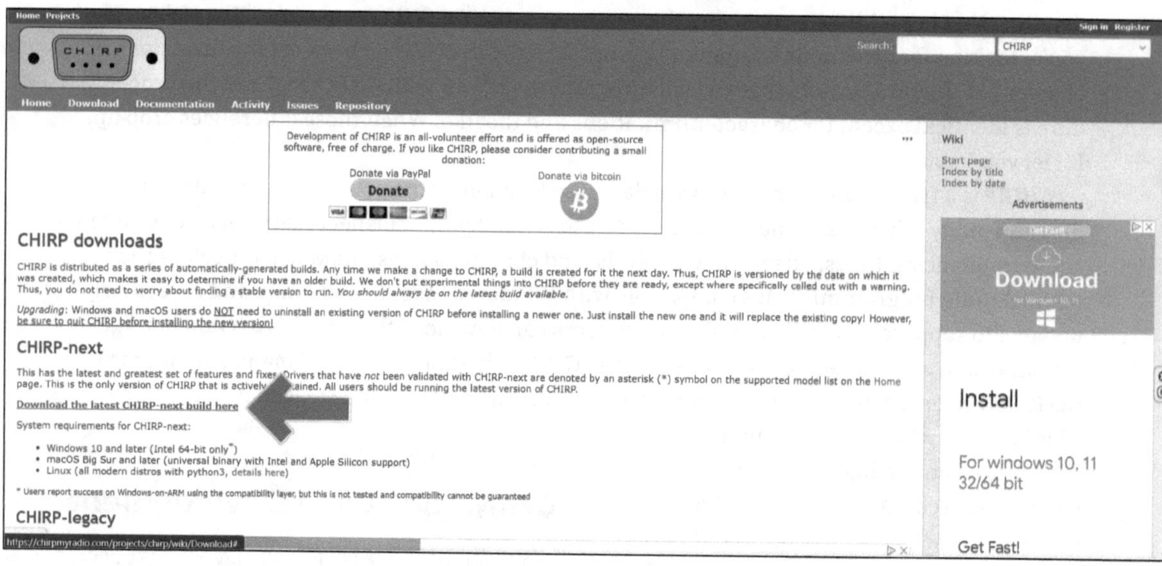

BaoFengTech supports the CHIRP project. If you are in the market for radios or accessories, please consider giving them your business!

If your virus scanner tells you that your download is a virus or malware, please see this page!

Name	Last modified	Size	Description
Parent Directory		-	
chirp-next-20250110.app.zip	2025-01-10 06:53	43M	MacOS Application
chirp-next-20250110-win64.zip	2025-01-10 06:53	23M	Windows standalone application
chirp-next-20250110-installer.exe	**2025-01-10 06:53**	**19M**	**Windows Installer (recommended)**
chirp-20250110.tar.gz	2025-01-10 06:53	2.2M	Source Tarball
chirp-20250110-py3-none-any.whl	2025-01-10 06:53	1.8M	Python 3 wheel
build_info/	2025-01-10 06:53	-	
SHA1SUM	2025-01-10 06:53	828	

STEP 2: INSTALL *CHIRP* ON A WINDOWS COMPUTER

1. Once downloaded, locate the .exe installer file in your Downloads folder and double-click it to start the installation.
2. The **Installation Wizard** pops up. Follow the directions to:
   - Accept the license agreement and click **Next.**
   - Choose the installation location and click **Next.**
   - Click **Install** to begin the installation.
3. Once the installation is complete, click **Finish** to exit the installer.

STEP 3: LAUNCH *CHIRP*

1. You can find *CHIRP* wherever you installed it. Double-click the *CHIRP* icon to launch the program. If you have trouble locating the program icon, the default install location is Local Disk > Program Files (x86) > Chirp. Locate the launcher **chirpwx.exe** and double click on it to start the program.
2. Connect your radio to the PC using a Baofeng official USB programming cable. If you have trouble locating the COM port your programming cable is connected to, you may need to download and install the drivers for your programming cable. For the Baofeng K-Plug programming cable, you can find the driver download in the **OTHER** section of the website: **www.baofengradio.com/pages/download.**

I have been using the RT Systems K4Y-USB cable with good success.

Rotate the volume control to halfway and ensure you are not on a channel that has traffic because their audio will cause errors.

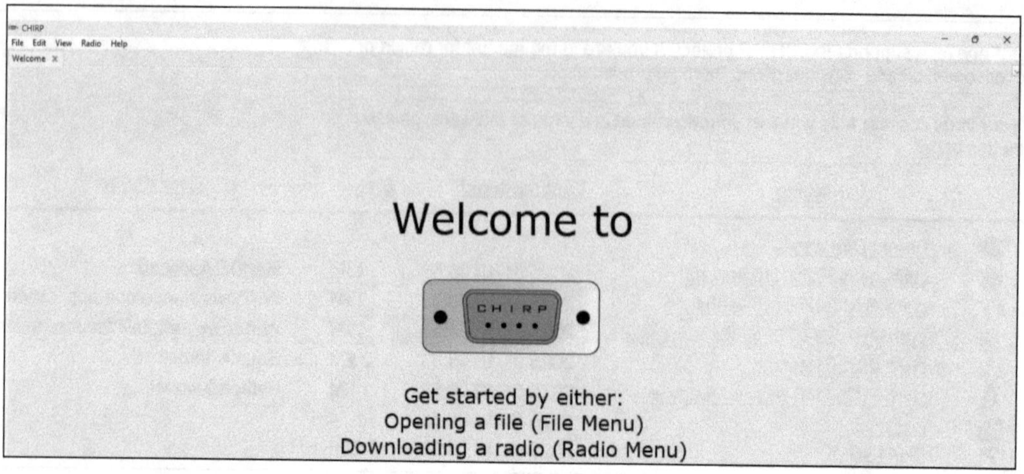

STEP 4: DOWNLOAD FROM RADIO

1. In *CHIRP,* click on the **Radio** dropdown in the upper left corner of the screen and select **Download from radio....**

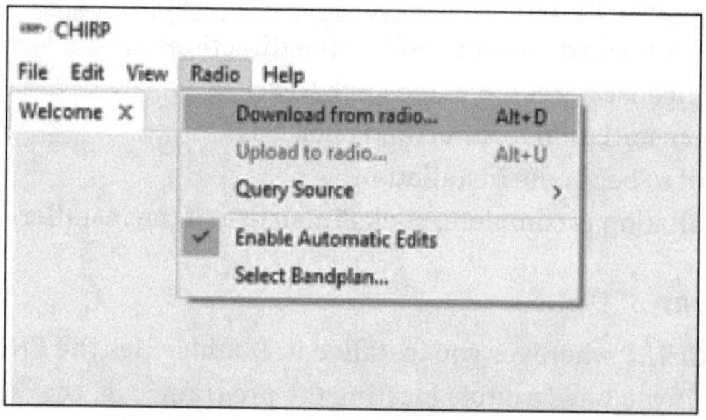

2. A pop-up window called **Communicate with radio** will open with three dropdowns lists: **Port, Vendor,** and **Model.**
   - In the **Port** dropdown, select the port that corresponds to the one you plugged the cable into.
   - In the **Vendor** dropdown, select **Baofeng.** After you've selected **Baofeng** a list of their models will populate the **Model** dropdown.
   - Select **UV-82.**

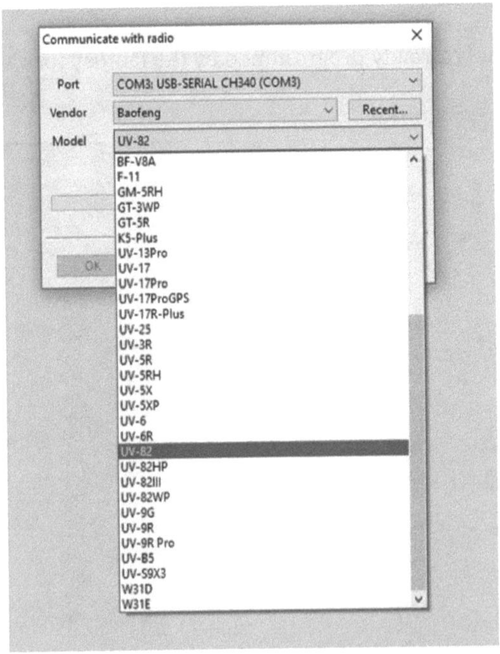

- Likely, a warning informing you that the model you selected "uses an experimental driver because of frequent changes to the radio's firmwear" will pop up. If you are not worried about seeing this message again, you can check the box next to **Do not prompt again for Baofeng UV-82?** To proceed with programming your radio, click **Yes** to dismiss the warning.
- Click **OK** on the **Communicate with radio** window to read frequencies and settings from your radio.
- Another dialog box called **Download Instructions** should pop up. Follow the directions on this dialog box, then click **OK** to start the download from your radio. The **Cloning from radio** status line fills in with green as the transfer moves toward completion.

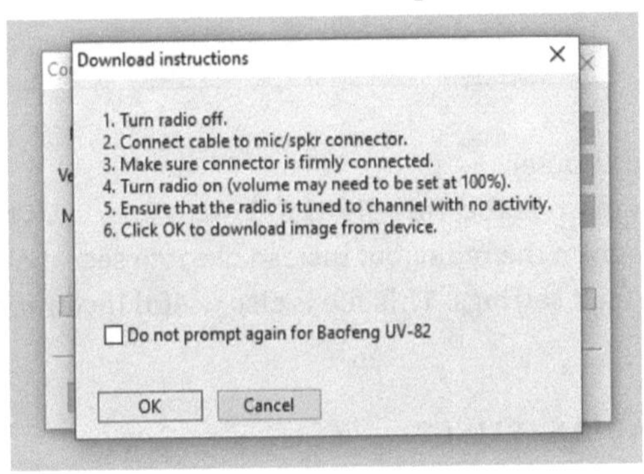

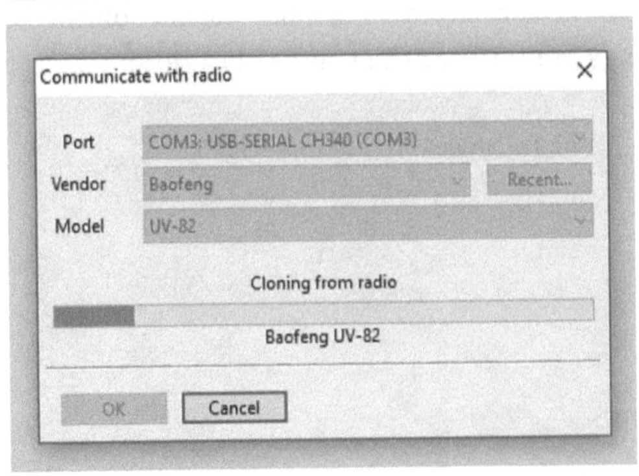

STEP 5: MAIN SCREEN

You should see the following 21 memory channels already programmed by the factory for quality control testing.

CHIRP (Baofeng_UV-82_20241230.img)

File Edit View Radio Help

Baofeng_UV-82_20241230.img* X

Memories Settings

	Frequency	Name	Tone Mode	Tone	Tone Squelch	DTCS	RX DTCS	DTCS Polarity	Cross Mode	Duplex	Offset/ TX Freq	Mode	Skip	Power
0														
1	452.125000		TSQL		69.3							FM		High
2	453.225000		TSQL		91.5							FM		High
3	454.325000		TSQL		136.5							FM		High
4	455.425000		TSQL		151.4							FM		High
5	456.525000		TSQL		192.8							FM		High
6	457.625000		TSQL		241.8							FM		High
7	458.725000		DTCS			025		NN				FM		High
8	459.825000		DTCS			134		NN				FM		High
9	461.925000		DTCS			274		NN				FM		High
10	462.225000		DTCS			346		NN				FM		High
11	463.325000		DTCS			503		NN				FM		High
12	464.425000		DTCS			073		RR				FM		High
13	465.525000		DTCS			703		RR				FM		High
14	402.225000											FM		High
15	437.425000											FM		High
16	479.975000											FM		High
17	138.550000											FM		High
18	157.650000											FM		High
19	172.750000											FM		High
20	438.500000											FM		High
21	155.700000											FM		High
22														

Save this file to your computer.

I like to use my call sign, so I label this file as follows: VE3IPS_UV82_20250114_OEM.img. I don't use this file to program the radio, but instead use it to see what frequencies are pre-programmed and the default settings. This file is also useful if you wish to restore the radio to its defaults.

STEP 6. CONFIGURE RADIO SETTINGS

On the left-hand side of the screen, there is a navigation column with options for viewing the **Memories** and **Settings** tabs. To see the standard radio settings from the factory, click on **Settings**.

The list of the settings is as follows:

- **Basic**
- **Advanced**
- **Other**
- **Work Mode**
- **DTMF**
- **Service settings**

BASIC SETTINGS MENU

The **Basic Settings** include a few things that can be used to personalize your radio, such as backlight LED color, timeout timer, and more.

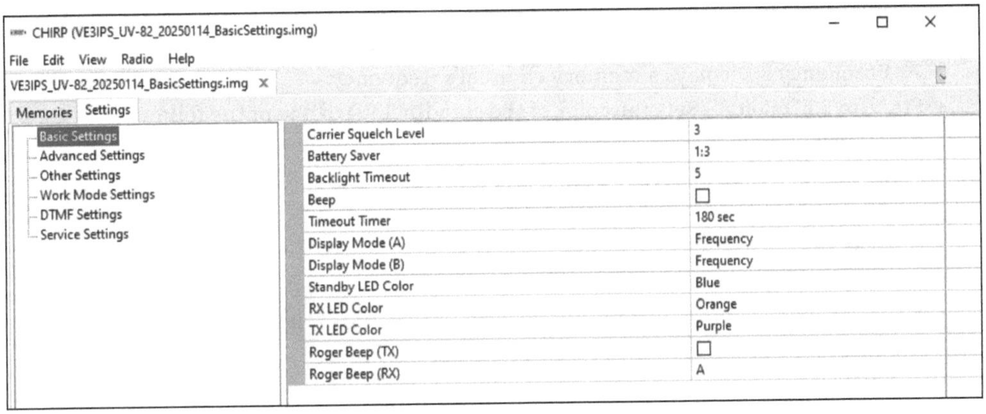

Here's a list of the settings available as **Basic Settings** with a few words of explanation.

- **Carrier Squelch Level** – This setting adjusts the squelch level of your received audio. Higher numbers are more restrictive (requiring a louder signal to "open up" the squelch). A pair of arrows on the right-hand side of this setting allow you to increase or decrease the squelch on a scale of **1 – 9.**
- **Battery Saver** – This setting changes how frequently the receiver is on and ready to receive a signal. The options are as follows:
  - **OFF** – The radio is always ready to receive.
  - **1:1** – The receiver is on and off an equal amount.
  - **1:2** – The receiver is on one time for every two times that it is off.
  - **1:3** – The receiver is on one time for every three times that it is off.
  - **1:4** – The receiver is on one time for every four times that it is off.
- **Backlight Timeout** – This sets the number of minutes the backlight will stay on while the radio is idle. A pair of arrows on the right-hand side of this setting allow you to increase or decrease the time on a scale of **1 – 24.**
- **Beep** – Toggles a beep on or off when a button is pressed. If **Voice** is on, the beep will not play when the voice speaks. It will only play on unvoiced buttons, such as **EXIT/AB** and **▲** / **▼**, are pressed.
- **Timeout Timer** – Prevents excessive transmission length. Can be set to any 15 second increment between **15** and **315** seconds using a dropdown.
- **Display Mode A** – Changes what the upper display shows using a dropdown with the following options:
  - **Channel** – Displays a memory channel's number.
  - **Name** – Displays a memory channel's **Name** field.

- **Frequency** – Displays a memory channel's frequency.

- **Display Mode B** – Changes what the lower display shows using a dropdown with the following options:
  - **Channel** – Displays a memory channel's number.
  - **Name** – Displays a memory channel's Name field.
  - **Frequency** – Displays a memory channel's frequency.
- **Standby LED Color** – Sets the color of the standby LED to one of the following using a dropdown:
  - **OFF**
  - **Blue**
  - **Orange**
  - **Purple**
- **RX LED Color** – Sets the color of the receive LED to one of the following using a dropdown:
  - **OFF**
  - **Blue**
  - **Orange**
  - **Purple**
- **TX LED Color** – Sets the color of the transmit LED to one of the following using a dropdown:
  - **OFF**
  - **Blue**
  - **Orange**
  - **Purple**
- **Roger Beep (TX)** – Transmits a beep when the PTT button is let go of. This setting can be annoying when used on the air or on a repeater, especially since most repeaters already transmit courtesy tones.
- **Roger Beep (RX)** – Plays a beep after a transmission is received on either the A or B frequency. In testing, the radio did not beep after receiving on either channel. The options available in the dropdown menu:
  - **OFF**
  - **A**
  - **B**

PROGRAMMING THE BASIC SETTINGS

The first setting I'll adjust is the backlight LED color.

1. First, click on the **Settings** tab.
2. Select **Basic Settings.**
3. Click on the cell to the right of **Standby LED Color** and select **Blue** from the dropdown.

Repeat the procedure above for the **RX LED Color** and **TX LED Color**, setting the **RX LED Color** to **Orange** and the **TX LED Color** to **Purple**.

Next, set the **Timeout Timer** to **180** seconds.

1. Click on the cell to the right of **Timeout Timer**.
2. From the dropdown, select **180 sec**.
3. Click on the **File** dropdown and select **Save As...**
4. Save the file using your naming convention, in my case I saved it as VE3IPS_UV-82_20250114_v1.img.

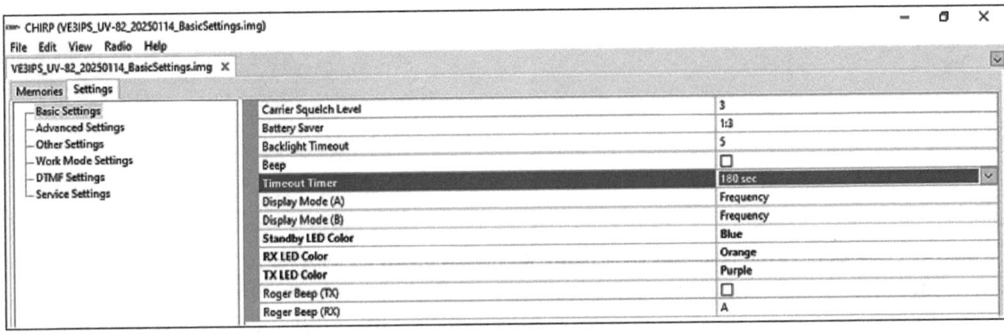

As I add and delete memories, I update the version. Remember to save your configuration file by clicking on the **File** dropdown and selecting **Save As...**

When I make changes, I toggle up the versions (v1, v2, v3, etc) instead of replacing an existing file in case I introduce errors and need to recover the last version.

STEP 7: UPLOAD TO RADIO

BASIC SETTINGS

Now it's time to upload the file with the adjusted settings to the radio.

1. In *CHIRP,* click on the **Radio** dropdown and select **Upload to Radio.**
   a. A pop-up window may appear asking whether you want to proceed with the upload. Click **Yes** to continue.

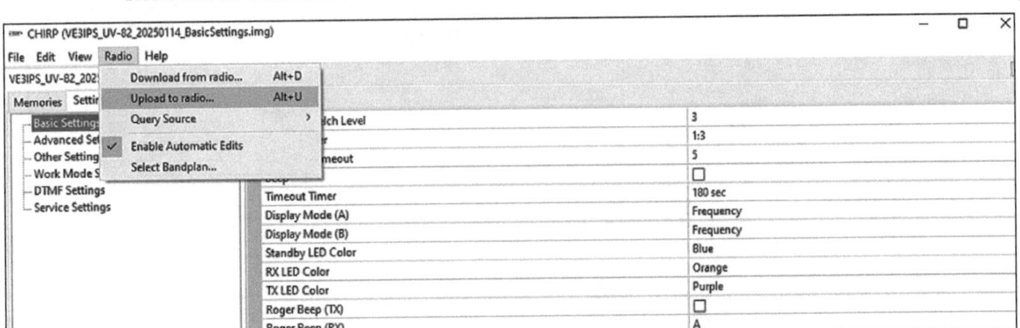

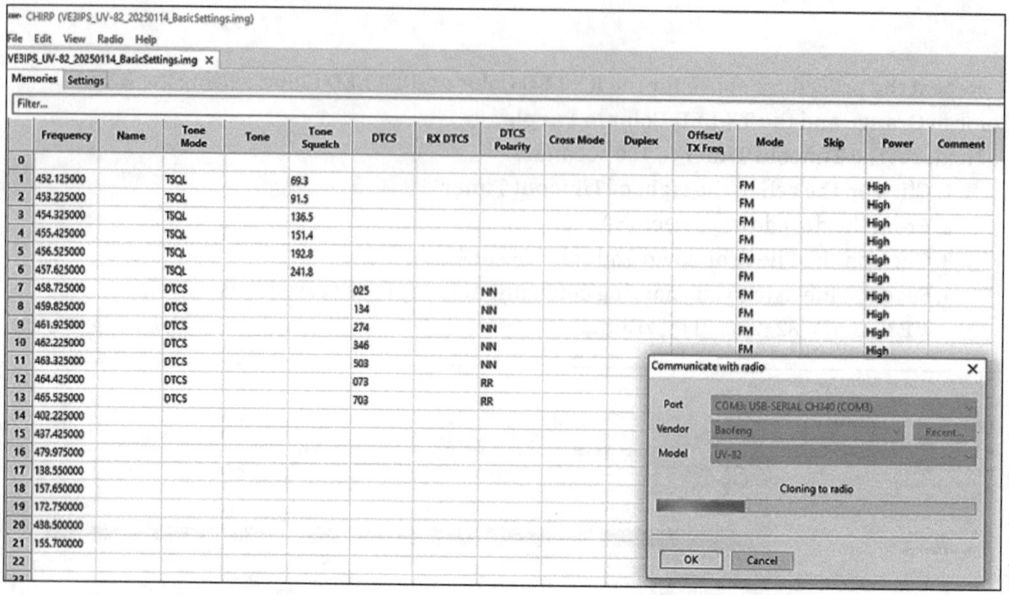

2. Click **OK** on the **Communicate with radio** window and the upload will start.
   a. The left indicator on the top of the radio will blink while the upload progresses. When it stops and the green on-screen indicator fills all the way in, the update is complete.
3. Unplug the programming cable and test the radio by checking to see that the LED colours are as you programmed. Transmit on your ham radio frequencies viewing the offsets in the display and checking to see if they are correct.

7.4.5 A Detour into the Remaining *CHIRP* Settings Tabs

ADVANCED SETTINGS

Plug the programming cable back in and let's look at the **Advanced Settings** tab.

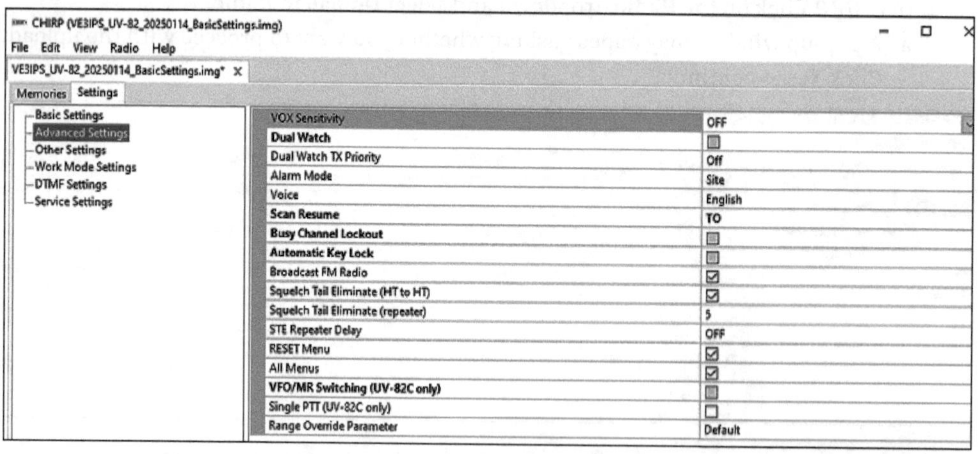

This tab allows you to adjust several more advanced settings, some of which we've already talked about. For example, you can set the **Alarm Mode**, which we've already adjusted to **Site**. You can also control whether the voice speaks in English or Chinese. Here's a list of the settings you can adjust and some basic information about them:

- **VOX Sensitivity** – this allows you to set whether the voice activated transmission is on or off and the level of its sensitivity.
- **Dual Watch** – toggles dual watch on or off.
- **Dual Watch TX Priority** – Let's you determine which frequency (**A**, **B**, or **Off**) to prioritize transmitting on. You can still hear both frequencies.
- **Alarm Mode** – Already discussed. Plays a local alarm (**Site**), transmits an emergency tone (**Tone**), or code (**Code**).
- **Voice** – Toggle the voice on or off and allows you to choose its language. Options in the dropdown menu are:
  - **Off**
  - **English**
  - **Chinese**
- **Scan Resume** – Sets how the scan feature works.
  - **TO** – Stops scanning when a transmission is detected and resumes after a preset amount of time.
  - **CO** – Stops scanning when a transmission is detected and resumes after the transmission goes silent.
  - **SE** – Stops scanning when a transmission is detected and does not resume until you manually restart the scan.
- **Busy Channel Lockout** – Toggle busy channel lockout (BCL) on or off. This prevents you from transmitting on a channel already in use. Turning this on can help prevent interference.
- **Automatic Key Lock** – Locks the keypad. Can be unlocked using the **#** key.
- **Broadcast FM Radio** – Enables or disables the broadcast radio band.
- **Squelch Tail Eliminate (HT to HT)** – Sends a 55 Hz tone to other Baofengs, which turns off their receivers momentarily. See sections 7.3.1 for a discussion of squelch and 7.6.11 for more discussion of squelch settings. Does not impact repeater operation.
- **Squelch Tail Eliminate (repeater)** – Sends a 55 Hz tone to the repeater, which generally does nothing and turns off your receiver for a moment. This is generally not a good idea because you will not know if you were transmitting into the repeater at the same time as someone else (called *doubling*) or if someone is replying immediately after your transmission. In general, I recommend leaving this setting off.
- **STE Repeater Delay** – Sets the amount of time your receiver is off (in milliseconds) after you finish your transmission. Range is from 100 to 1,000 ms. In general, leave this setting on OFF for the same reasons given in the previous item.
- **RESET Menu** –Disables the **RESET** menu when off.

- **All Menus** – Disables all menus when off. Allows only for channel mode.
- **VFO/MR Switching (UV-82C only)** – Turns on or off the ability to switch between VFO and memory modes on the UV-82C.
- **Single PTT (UV-82C only)** – Enables or disables transmit only on the selected channel for UV-82Cs.
- **Range Override Parameter** – Not an alterable function. The only option is to use the default setting.

At this point in the programming exercise there is no need to make changes, though you could if you would like. For example, if you'd like to turn Dual Watch ON, you'll be able to listen for contacts on multiple frequencies.

Let's go to the **Other Settings** tab before disengaging the programming cable.

OTHER SETTINGS

The **Other Settings** tab allows you to program power-on messages (which I'll walk you through shortly), set the VHF and UHF band limits, and enable or disable transmit on VHF and UHF. Here's a full list of the options you should see in the **Other Settings** tab.

- **Firmware Message 1** – This displays the firmware version installed on the radio. This is not adjustable in the version of *CHIRP* available as of this writing. If you search for how to verify your firmware number on YouTube and other sources you may hear that holding 3 on powerup will display the radio's firmware version. However, back in 2013 Baofeng stopped writing the firmware versions to their radios' screens. For example, holding 3 while powering on my radio simply displays VER NUV82. This is just a generic firmware number many UV-82s display.
- **Firmware Message 2** – This is not an adjustable setting.
- **6+ Power-On Message 1** – This is not an adjustable setting. This should be what displays when you hold 6 while powering on your radio. However, in my case, if I hold 6 the radio does not display number 151123H, as shown in *CHIRP* and in the image below, but instead displays 190923M. These are both likely dates relating to hardware, manufacture, or something similar.
- **6+ Power-On Message 2** – This is not an adjustable setting.
- **Power-On Message 1** – This field allows you to change the message that appears on the radio's startup screen.
- **Power-On Message 2** – This field allows you to add an additional message to the radio's startup screen.
- **VHF Lower Limit (MHz)** – This field allows you to adjust the lowest frequency the radio will receive and transmit on in the VHF band up or down a few MHz. The parameters of the radio's hardware will restrict how high or low this will realistically adjust.
- **VHF Upper Limit (MHz)** – This field allows you to adjust the highest frequency the radio will receive and transmit on in the VHF band up or down a few MHz. The parameters of the radio's hardware will restrict how high or low this will realistically adjust.

- **VHF TX Enabled** – Enables or disables transmissions on the VHF band.
- **UHF Lower Limit (MHz)** – This field allows you to adjust the lowest frequency the radio will receive and transmit on in the UHF band up or down a few MHz. The parameters of the radio's hardware will restrict how high or low this will realistically adjust.
- **UHF Upper Limit (MHz)** – This field allows you to adjust the highest frequency the radio will receive and transmit on in the UHF band up or down a few MHz. The parameters of the radio's hardware will restric how high or low this will realistically adjust.
- **UHF TX Enabled** – Enables or disables transmissions on the UHF band.
- **Power-on Message** – Allows you to select one of two settings.
  - **Message** – Displays a two-line power-on message.
  - **Full** – Displays the entire LCD screen.

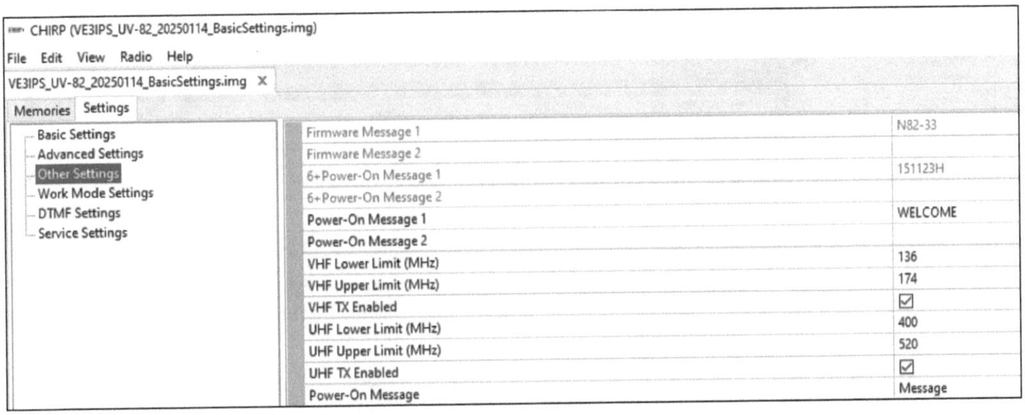

In this settings group, I would leave most of the settings at their default, except for Power-On Message 1, which I like to change to my call sign, VE3IPS. Other users also like to utilize the second line to show their name, which can be useful, however the UV-82's screen is limited to displaying seven characters, putting many names like mine, LEONARDELLI, outside the character limit. Adjusting this setting could be useful if you misplace your radio, lend it to a friend, or are using it in a situation where it might be mixed up with other radios, such as during public service or emergency communications, so let's go ahead and implement this change.

1. First click on the box that says **WELCOME** beside **Power-On Message 1**.
2. Delete **WELCOME** and input your call sign. In my case, I'll input **VE3IPS.** Note that *CHIRP* only accepts capital letters, numbers 0 – 9 and most common keyboard symbols, including @, #, &, (,), -, and /.
3. Now, click on the **File** dropdown and select **Save As...**
4. Adjust the file name according to your file naming convention. Mine is VE3IPS_UV-82_20250114_BasicSettings_v2.img. Notice that I've moved the version up to 2 from 1.
5. Click **Save.**
6. After saving your file, click on the Radio dropdown and select **Upload to radio.** The window **Communicate with radio** should pop up, prepopulated with the port you've been using, the vendor, and model number.

7. Click **OK.** If the **Upload instructions** window pops up, follow the directions, and click **OK** to continue.
    a. The left indicator on the top of the radio will blink while the upload progresses. When it stops and the green on-screen indicator fills all the way in, the update is complete.
8. Turn off the radio and unplug the programming cable from it.
9. Power on your radio and your call sign should display where the word **WELCOME** previously did.

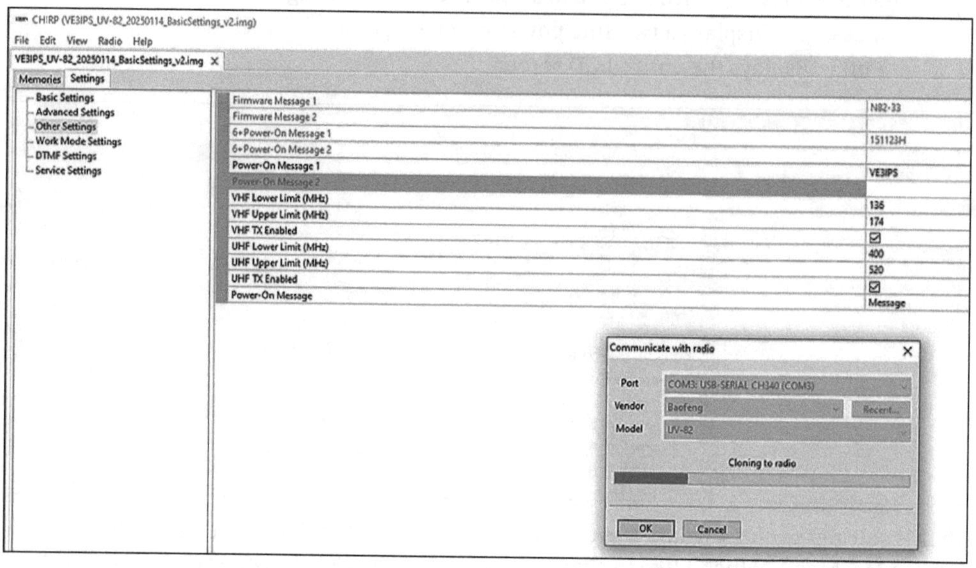

WORK MODE SETTINGS

The **Work Mode Settings** contain many useful items. All of them, however, are easily adjusted manually using the procedures detailed in chapter 6. Because of that, I won't go into any detail on them here, but I will provide the list of settings.

- **Display** – This dropdown allows you to select **A** or **B.**
- **VFO/MR Mode** – Switches between **VFO** and **Memory** modes.
- **Keypad Lock** – Locks the keypad. Hold the # key to toggle manually.
- **MR A Channel** – Sets the memory channel displayed on line **A** when in **Memory Mode.** Choose from **0 – 127** in the dropdown.
- **MR B Channel** – Sets the memory channel displayed when in **Memory Mode.** Choose from **0 – 127** in the dropdown.
- **VFO A Frequency** – Sets the frequency displayed in line **A.** Easily programmed manually.
- **VFO B Frequency** – Sets the frequency displayed in line **B.** Easily programmed manually.

- **VFO A Shift** – Select **Off**, **+**, or **-** from the dropdown.
- **VFO B Shift** – Select **Off**, **+**, or **-** from the dropdown.
- **VFO A Offset (0.0-999.999)** – Sets the offset in MHz.
- **VFO B Offset (0.0-999.999)** – Sets the offset in MHz.
- **VFO A Power** – Select **High** or **Low** from the dropdown.
- **VFO B Power** – Select **High** or **Low** from the dropdown.
- **VFO A Bandwidth** – Select **Wide** or **Narrow** from the dropdown.
- **VFO B Bandwidth** – Select **Wide** or **Narrow** from the dropdown.
- **VFO A PTT-ID** – Select a value of **1 – 15** from the dropdown.
- **VFO B PTT-ID** – Select a value of **1 – 15** from the dropdown.
- **VFO A Tuning Step** – Select a value from the dropdown:
  - **2.5**
  - **5.0**
  - **6.25**
  - **10.0**
  - **12.5**
  - **20.0**
  - **25.0**
  - **50.0**
- **VFO B Tuning Step** – Select a value from the dropdown:
  - **2.5**
  - **5.0**
  - **6.25**
  - **10.0**
  - **12.5**
  - **20.0**
  - **25.0**
  - **50.0**

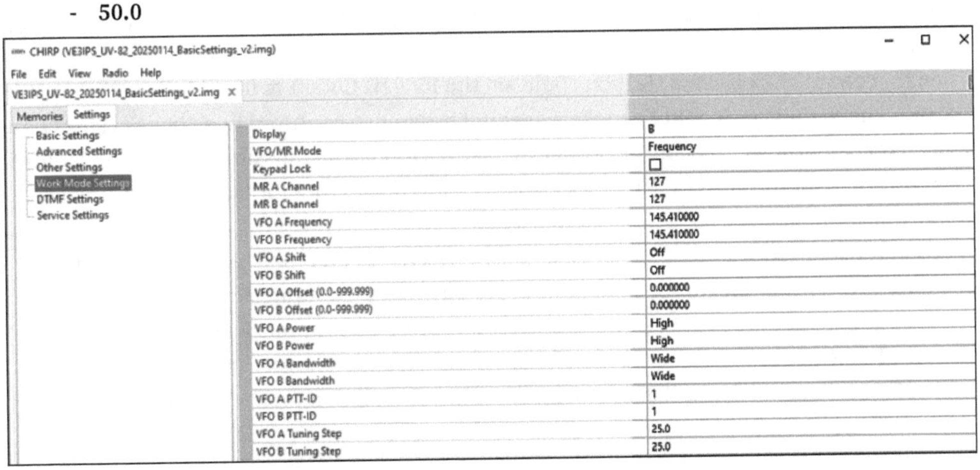

DUALTONE MULTI-FREQUENCY (DTMF)

The DTMF (Dual-Tone Multi-Frequency) keypad on many ham radios serves multiple functions, and while its role has evolved over time, it remains a valuable tool for radio operators. The **0 - 9** keypad uses distinct tones, just like on a telephone, to pass DTMF commands over the air.

In the past, many repeaters were linked to landline telephone systems, allowing hams to make phone calls from their radios — a system known as auto patch. By using the DTMF keypad, hams could dial phone numbers through the repeater. While this use has largely disappeared with the rise of cell phones, it was once a vital function for mobile hams. There may still be auto patch equipped repeaters in your area.

Repeater networks also use DTMF or "touch tones" to access repeaters through linking commands. Many modern repeater systems allow over-the-air (OTA) programming and control via DTMF tones. You can make changes to the repeater settings remotely using a password.

Also, there are DTMF controller boards that use DTMF commands to activate relays that can be used for station control functions.

ALLSTARLINK, IRLP, AND *ECHOLINK®* ACCESS

AllStarLink, the *Internet Radio Linking Project (IRLP)*, and *EchoLink* are networks of radio stations that use the voice-over internet protocol (VoIP) to digitize sound and pass it through the internet. All these networks utilize DTMF tones for node control. Operators can:

- Connect and disconnect nodes by entering specific DTMF codes.
- Link to distant repeaters and networks, enabling global communications.
- Control *AllStar* node functions, such as adjusting audio levels or changing configurations remotely. For example, the "1" DTMF tone is used to disconnect a link.

While auto patch has declined, the DTMF keypad remains essential for modern ham radio applications.

DTMF SETTINGS

The settings in the **DTMF Settings** section allow the user to manually enter up to a five-digit PTT ID code that can be associated with memory channels and sent whenever the radio transmits on that channel. For example, you could set the PTT ID Code 1 to 00101 (all values must be 00101 or higher), then assign that code to a specific frequency captured in a channel (say channel 1). That value will be transmitted either before or after a transmission, depending on how you set settings elsewhere. The PTT ID settings are intended more for business use and are generally beyond the needs of a ham radio user.

For reference, here are the settings available in this menu and their possible adjustments:

- **PTT ID Code 1** – Set to 20202. Can be set to any five-digit numerical value.
- **PTT ID Code 2** – Set to 11111. Can be set to any five-digit numerical value.
- **PTT ID Code 3** – Set to 22222. Can be set to any five-digit numerical value.
- **PTT ID Code 4** – Set to 33333. Can be set to any five-digit numerical value.

- **PTT ID Code 5** – Set to 44444. Can be set to any five-digit numerical value.
- **PTT ID Code 6** – Set to 55555. Can be set to any five-digit numerical value.
- **PTT ID Code 7** – Set to 66666. Can be set to any five-digit numerical value.
- **PTT ID Code 8** – Set to 77777. Can be set to any five-digit numerical value.
- **PTT ID Code 9** – Set to 88888. Can be set to any five-digit numerical value.
- **PTT ID Code 10** – Set to 99999. Can be set to any five-digit numerical value.
- **PTT ID Code 11** – Set to 12121. Can be set to any five-digit numerical value.
- **PTT ID Code 12** – Set to 34343. Can be set to any five-digit numerical value.
- **PTT ID Code 13** – Set to 56565. Can be set to any five-digit numerical value.
- **PTT ID Code 14** – Set to 78787. Can be set to any five-digit numerical value.
- **PTT ID Code 15** – Set to 30303. Can be set to any five-digit numerical value.
- **ANI Code** – Automatic Number Identification (ANI). Defaults to **80808**, but can be set to any value greater than 00101.
- **ANI ID** – Sets when the ANI Code is transmitted. Available options are:
  - **Off**
  - **BOT** – Sends ANI ID at the beginning of the transmission.
  - **EOT** – Sends ANI ID at the end of the transmission.
  - **Both** – Sends ANI ID at the beginning and end of the transmission.
- **Alarm Code** – Sets the code transmitted when the alarm is set to **Code.** If another Baofeng radio is within receive distance and receives a code that matches the one programmed into their radio, their radio will play its alarm. If you are in a small group, say out hiking, and you all set your alarm code to a specific number, if members get separated someone can send this code to let the entire group know that there's an emergency. You can then communicate to figure out what is going on and who needs help.
- **DTMF Sidetone** – This allows you to hear the tones being transmitted by your radio. If you are transmitting DTMF or ANI codes, it's a good idea to be able to hear them yourself. The options you can select from the dropdown are:
  - **OFF** – Turns off the sidetones.
  - **DT-ST** – Plays the DTMF sidetones.
  - **ANI-ST** – Plays the ANI ID sidetones.
- **DT+ANI** – Plays both the DTMF and ANI ID sidetones.
- **DTMF Speed (on)** – Sets the amount of time before the DTMF tones are transmitted.
- **DTMF Speed (off)** – Sets the amount of time used as spacing between DTMF tones if multiple tones are transmitted.
- **PTT ID Delay** – Sets the amount of time that elapses between when the **PTT** button is pressed and when the **PTT ID** is transmitted. Values range from **0** to **50** ms and can be set using the arrow in the box.

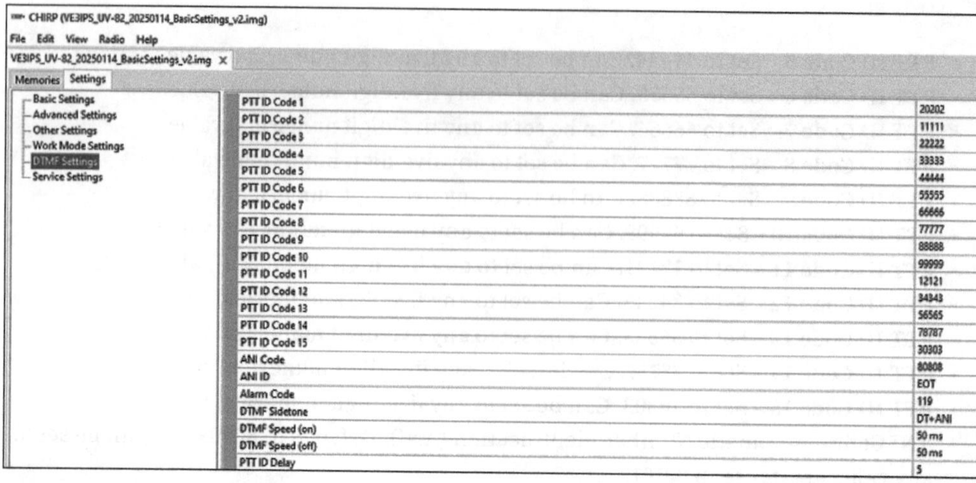

SERVICE SETTINGS

These settings allow you to adjust the squelch levels for both the VHF and UHF bands. The factory presets move the squelch up by one increment for each level. **VHF Squelch 1** starts at **20**, **VHF Squelch 2** moves to **21**, and so on for squelches 1 – 9. The UHF squelches are set up the same way, but they start at **16** for squelch 1 and go to **24** for squelch 9. The range of squelch is much greater than the factory settings would suggest. The upper squelch limit is 130, which will make it practically impossible for any transmission to open the squelch, and the squelch goes down to 0 (which, to be clear, is what the VHF and **UHF Squelch 0**s are set to). To adjust the level for a given squelch setting, click on the cell containing the squelch level you would like to adjust and scroll through the numbers until you arrive at the number you want.

The best way to figure out which squelch level is appropriate for each setting is to try them out and determine what's best for the type of operating you do. Because this setting tab is so uniform, I won't reproduce it in text, but will instead rely upon the screenshot to list out the various options.

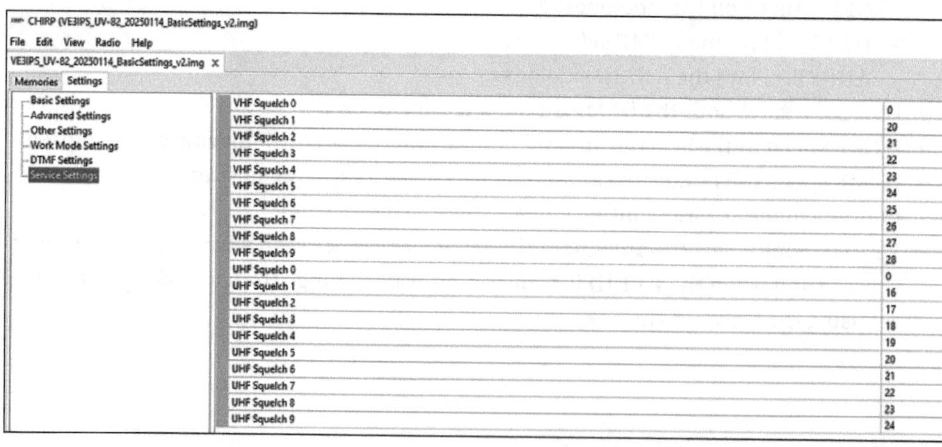

7.4.6 The Memories Spreadsheet

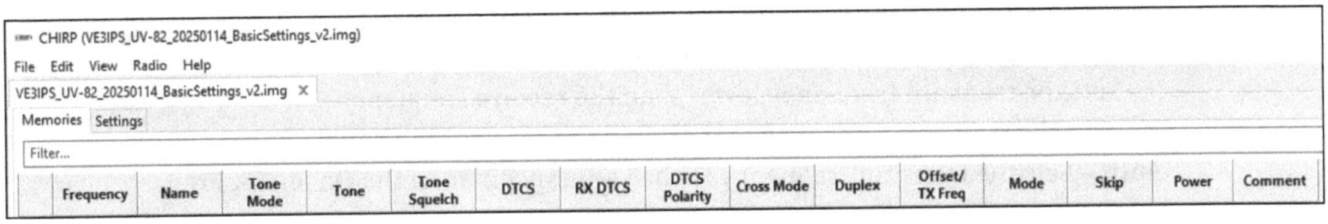

STEP 8: PROGRAMMING MEMORY ENTRIES

If you are continuing the procedure from the last section, you should have a blank memory spreadsheet in front of you. If you're picking up later, open your v2 file and click on the **Memories** tab.

Across the top of the spreadsheet, you should see a series of columns. Below I've listed them with brief explanations of the column heading and its possible values where appropriate.

- **Frequency** – Sets the receive frequency in MHz.
- **Name** – The name you have assigned to the entry.
- **Tone Mode** – Select from:
  - **Tone** – Tells the transceiver to send a CTCSS tone.
  - **TSQL** – Sets a single CTCSS tone for both transmit and receive.
  - **DTCS** – Tells the transceiver to send Digital Tone Coded Squelch. This is similar to CTCSS, but a digital mode.
  - **Cross** – Sends both CTCSS and DTCSS tones or multiple of either. Very rarely used.
- **Tone** – Select which CTCSS tone to send.
- **Tone Squelch** – Only receives signals that are transmitting the specified tone code.
- **DTCS** – Sets the value of the DTCS (DCS) code.
- **RX DTCS** – Same as Tone Squelch, but uses DTCS codes.
- **DTCS Polarity** – Sets the polarity of the DTCS tones to either normal or reverse. The available options are:
  - **NN** – Normal Normal, meaning both transmit and receive polarity are normal.
  - **RN** – Reverse Normal, meaning transmit is reverse polarity and receive is normal polarity.
  - **NR** – Normal Reverse, meaning transmit is normal polarity and receive is reverse polarity.
  - **RR** – Reverse Reverse, meaning transmit is reverse polarity and receive is reverse polarity.
- **Cross Mode** – Specifies the order of tones sent. The options are:
  - **Tone -> Tone** – One tone is used for transmit and a different tone is used for receive.
  - **Tone - > DTCS** – A tone is used for transmit and a DTCS for receive.
  - **DTCS - > Tone** – A DTCS is used for transmit and a tone is used for receive.
  - - > **Tone** – Not tone is transmitted but one is used for receive.
  - - > **DTCS** – No DTCS is transmitted but one is used for receive.

- **DTCS - >** – DTCS is transmitted and carrier squelch is used for receive.
- **DTCS - > DTCS** – One DTCS is used for transmit and a different DTCS is used for receive.

- **Duplex** – Determines the transmit frequency. The options are:
  - **Blank** – Receive and transmit frequencies are the same. Simplex.
  - - – Sets the transmit frequency lower than the receive frequency.
  - + – Sets the transmit frequency higher than the receive frequency.
  - **split** – Sets the transmit frequency to the value specified in the Offset/TX Freq column.
  - **off** – Prevents transmission. Only allows the radio to receive.
- **Offset/TX Freq** – Adjusts the transmit frequency up or down by the specified amount. For example, a standard offset in VHF is .600 MHz. (600 kHz). If Duplex is set to + the transmit frequency will be .600 MHz higher than the receive frequency. If Duplex is set to split, then you should enter a specific frequency in this box for the radio to transmit on.
- **Mode** – Sets the FM bandwidth. Options are:
  - **FM** – Frequency modulation. The setting used for amateur radio (Part 97) transmissions. The FM transmit bandwidth is set to 5 kHz.
  - **NFM** – Narrow frequency modulation. The setting used for other radio services. The FM transmit bandwidth is set to 2.5 kHz.
- **Skip** – When scanning, tells the radio to skip a specific channel when S is selected.
- **Power** – Sets the transmit power level. Options are:
  - **High** – Transmits at maximum power (either 8 W or 4 W depending on the model).
  - **Low** – Transmits at 1 W.
- **Comment** – Space for additional user comments.

DELETING MEMORY ENTRIES

Before adding memory entries to your radio with *CHIRP*, it's important that you know how to clear preexisting memories. The UV-82 comes with 21 factory memories preloaded. These are not likely to be useful to you because they might as well be random frequencies.

So that you don't lose the settings you just changed in v2, update the name of the file to v3 after deleting the 21 factory-set memories. By the time you're done with this next section, you should have four different *CHIRP* files saved on your computer. The first should have your changes to the **Basic Settings.** The next file, v2, has any changes you may have made to the **Advanced** and **Other** settings, and v3 should include all of the changes to the settings and a blank memory spreadsheet. The final save, v4, should include the five demonstration memory entries. Let's dive into deleting memory entries using *CHIRP.*

1. Click on the **Memories** tab to display the list of 21 factory-set memories.
2. To select all 21 lines of memory, click on the row number **1** beside the first **Frequency** entry. Then hold the **Shift ⇧** key and click on the row number **21** beside the final Frequency entry. This should select all 21 memories and all of their associated fields. Before deleting, notice that row 0 is blank. When you input your memory entries, you

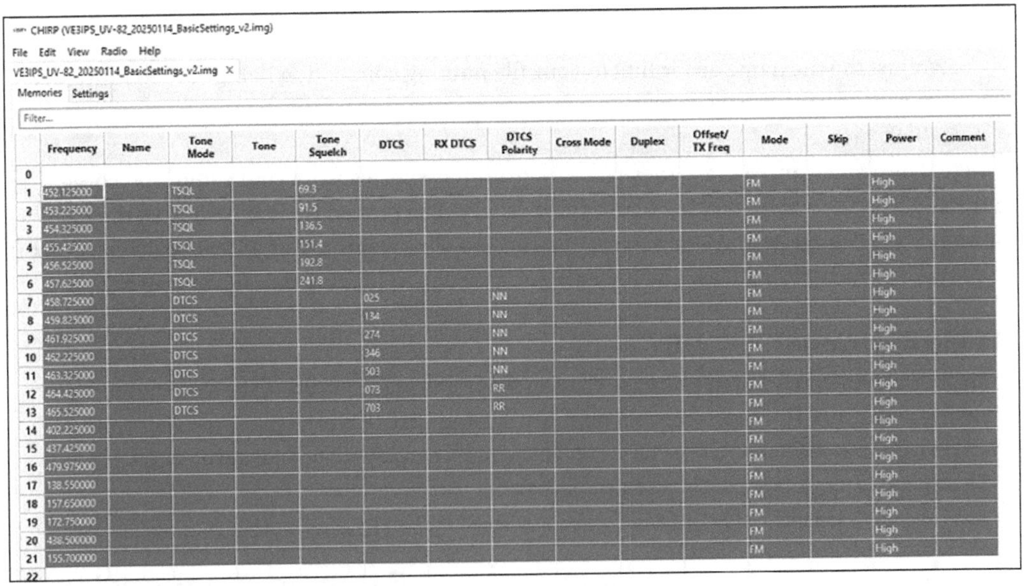

	Frequency	Name	Tone Mode	Tone	Tone Squelch	DTCS	RX DTCS	DTCS Polarity	Cross Mode	Duplex	Offset/ TX Freq	Mode	Skip	Power	Comment
0															
1	452.125000		TSQL		69.3							FM		High	
2	453.225000		TSQL		91.5							FM		High	
3	454.325000		TSQL		136.5							FM		High	
4	455.425000		TSQL		151.4							FM		High	
5	456.525000		TSQL		192.8							FM		High	
6	457.625000		TSQL		241.8							FM		High	
7	458.725000		DTCS			025		NN				FM		High	
8	459.825000		DTCS			134		NN				FM		High	
9	461.925000		DTCS			274		NN				FM		High	
10	462.225000		DTCS			346		NN				FM		High	
11	463.325000		DTCS			503		NN				FM		High	
12	464.425000		DTCS			073		RR				FM		High	
13	465.525000		DTCS			703		RR				FM		High	
14	402.225000											FM		High	
15	437.425000											FM		High	
16	479.975000											FM		High	
17	138.550000											FM		High	
18	157.650000											FM		High	
19	172.750000											FM		High	
20	438.500000											FM		High	
21	155.700000											FM		High	
22															

will need to leave row **0** blank too. Otherwise, your memory entries will start with the second row rather than the first. Enter the frequencies and attributes as per our list. The **NAME** field can accommodate 7 characters so you may need to be a little creative in your naming convention.

3. The **Backspace** ⇦ or **Delete** key only deletes single characters from inside a cell. To delete all of the entries at once you must right click o the spreadsheet and hover over **Delete**, then click on **21 memories.** This clears all of the selected memory cells.

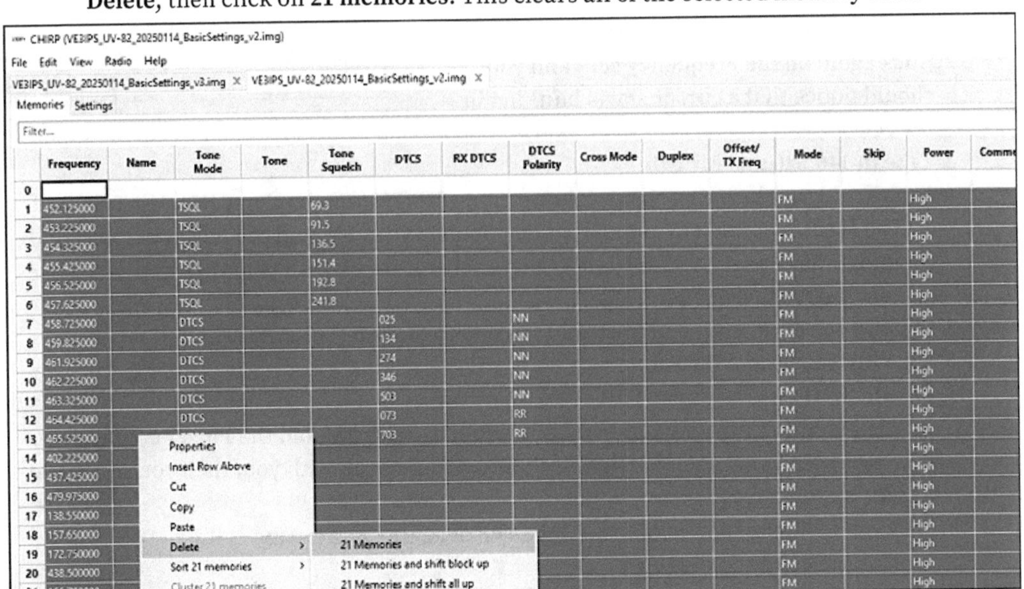

	Frequency	Name	Tone Mode	Tone	Tone Squelch	DTCS	RX DTCS	DTCS Polarity	Cross Mode	Duplex	Offset/ TX Freq	Mode	Skip	Power	Comm
0															
1	452.125000		TSQL		69.3							FM		High	
2	453.225000		TSQL		91.5							FM		High	
3	454.325000		TSQL		136.5							FM		High	
4	455.425000		TSQL		151.4							FM		High	
5	456.525000		TSQL		192.8							FM		High	
6	457.625000		TSQL		241.8							FM		High	
7	458.725000		DTCS			025		NN				FM		High	
8	459.825000		DTCS			134		NN				FM		High	
9	461.925000		DTCS			274		NN				FM		High	
10	462.225000		DTCS			346		NN				FM		High	
11	463.325000		DTCS			503		NN				FM		High	
12	464.425000		DTCS			073		RR				FM		High	
13	465.525000					703		RR				FM		High	
14	402.225000											FM		High	
15	437.425000											FM		High	
16	479.975000											FM		High	
17	138.550000											FM		High	
18	157.650000											FM		High	
19	172.750000											FM		High	
20	438.500000											FM		High	
21	155.700000											FM		High	

4. Click the **File** dropdown and select **Save as...**
5. Adjust the file name according to your file naming convention being sure to move the number from v2 to v3.
6. Click **Save.**

Note that if you want to erase these factory-preset memories from your radio, you could go through the steps to upload the blank spreadsheet. The procedure is the same as the one I showed you to change settings on your radio. Instead of rehashing that, I'll move directly on to programming memory entries, and I'll walk you through uploading that file to your radio.

TYPING IN MEMORY ENTRIES

To program memory entries, I'll use the same five channels I used in the last chapter. The following procedure for programming memories using CHIRP works the same way on the UV-5R except that radio has a six-character limit on its display instead of seven.

The memory channels I'll use as examples are:

- Memory 1: 146.520 Simplex
- Memory 2: 446.000 Simplex
- Memory 3: 162.400 weather channel
- Memory 4: W1AW VHF repeater 145.450, + 600, 127.3 Hz
- Memory 5: W1OKY UHF repeater 443.050, +5.0, 100 Hz

Example Frequencies

These frequencies are just examples. If there are other frequencies you would like to program, feel free to substitute them. You'll want a simplex frequency or two (the national calling frequency on 2 meters is 146.520, for example), a repeater or two with its access codes and offsets, and perhaps a local NOAA or Weatheradio Canada station.

The following directions will walk you through entering the five example frequencies using *CHIRP.*

1. Click on the **Frequency** cell in row **1.** The program will select the entire row.
2. Click again on the Frequency cell, and you should notice that a cursor starts blinking in the cell.
3. Type in **146.520** and hit **Tab.**
4. You should now be in the **Name** column. Type **SIMPV** (for simplex VHF). Hit **Tab.**
   - The **Mode** column will automatically populate with **FM** and the **Power** column will automatically set itself to **High.** If these setting are ok with you, then you can move onto the next frequency.
5. Click on the Frequency cell in row 2 then click again on the same cell.
6. Type in **446.000** and hit **Tab.**
7. Type in **SIMPU** (for simplex UHF) and press **Tab.**
   - The **Mode** column will automatically populate with **FM** and the **Power** column will automatically set itself to **High.** If these setting are ok with you, then you can move onto the next frequency.
8. Click on the Frequency cell in row 3 then click again on the same cell.
9. Type in **162.400** and press **Tab.**

10. You should now be in the **Name** column. Type **WX** (for weather).
11. Press **Tab** eight times to land on the cell in the **Duplex** column. Click on it once, pause, and then click again to pull up the drop-down menu.
12. Select **off**. This will prevent you from transmitting on the weather channel and causing needless interference.
13. Hit the **Home** key then press **Enter**. This should select cell **4** in the **Frequency** column.
14. Type in **145.450** and press **Tab**.
    - The **Mode** and **Power** columns will automatically populate.
15. Type **W1AWV** (for W1AW VHF repeater). Press **Tab.**
16. Select **Tone** from the **Tone Mode** drop-down. Press **Tab.**
    - The **Tone, Duplex**, and **Offset** columns will automatically populate.
17. Select **127.3** from the **Tone** dropdown. Press **Tab** six times to select the cell **4** in the **Duplex** column.
18. Select + from the dropdown.
19. Now, try programming the information for the W1OKY repeater on your own. If you run into trouble, consult steps 12 through 17 in the list above.

Don't Transmit on Public Service Frequencies

It's important not to transmit on channels that you don't want to or aren't permitted to transmit on. The NOAA Weather Radio frequency in this list makes a good example. 162.400 MHz is outside the ham bands. To make sure you don't transmit on it you can store the frequency in a memory channel and the set the **DUPLEX** field in the Memories tab in *CHIRP* to off. You could also turn on the busy channel transmit lock. Since weather broadcasts are continuous, you shouldn't be able to accidentally transmit on them. For more information on the Memories tab, see section 7.4.3 Programming the UV-82 or the UV-5R Using *CHIRP* and Step 8, on the Memories Spreadsheet. This is a handy feature for any frequency you don't want to transmit on. Setting frequencies up as channels gives you more control over a variety of features. If you listen on other public service frequencies, it's worth the time to set them up as channels you can use in the radio's memory and prevent transmission on them. Interference on these channels can cause real harm, not to mention it's illegal.

Once you've programmed the five memory channels, go ahead and save the file as v4 and plug the programming cable into the radio. Click on the **Radio** dropdown, select **Upload to radio**, and follow the steps to start the cloning process.

Turn the radio OFF and then ON again and check that the memory channels are populated as programmed. If you used local repeaters in your test, you can test them as well as the simplex frequencies you programmed by doing some test transmissions.

Before leaving the subject of programming your radio with *CHIRP*, it behooves me to walk through a quick way to program many local repeaters at once. The website repeaterbook.com contains tens of thousands of local repeaters with all of their information. By connecting *CHIRP* to RepeaterBook, you can quickly create a list of useful repeaters for just about anywhere in the United States and Canada.

STEP 9. IMPORTING REPEATERBOOK ENTRIES TO *CHIRP*

Importing RepeaterBook entries into *CHIRP* saves time and effort. It's especially useful when you're traveling, and you want to know what the local repeaters are. If you're new to ham radio, it's also a great way to find out what's happening in your area. RepeaterBook allows you to sort and filter data on repeaters in a number of different ways. You can view statewide data for all bands, or you can specify a particular band (or bands) you want to see. You can also limit the types of repeaters based on their affiliations with various groups like the Amateur Radio Emergency Service (ARES®) or SKYWARN®, or you can view all of the repeaters in a linked system. Here's the full list of filters:

- **Band**
- **Mode**
- **Feature**
- **Linked Systems**
- **Emergency Service Repeaters**
- **Repeaters recommended for use by nearby highway**
- **Repeaters by nearest city/town**
- **Repeaters by county**
- **Repeaters updated recently (days)**
- **Repeaters reviewed recently (days)**
- **Repeaters needing re-verification**

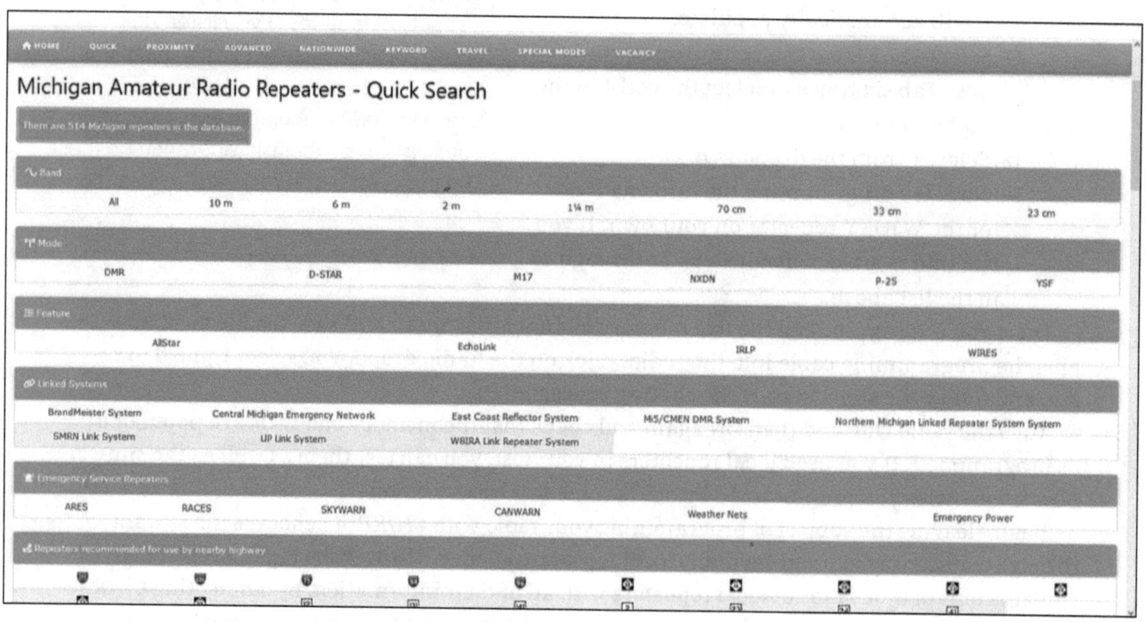

To pull in this wealth of information, you don't even have to visit the RepeaterBook website. Instead, you can pull it directly into *CHIRP*. Here's how to do it.

1. From the menu at the top of the screen, click on the **Radio** dropdown.
   - Hover over **Query Source.**
   - Select **Repeater Book.**

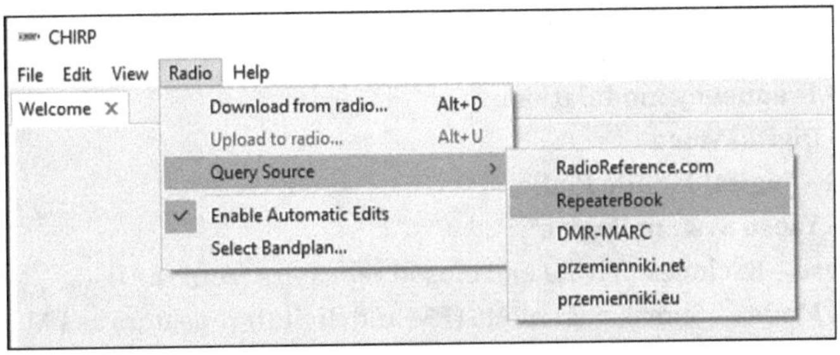

2. A window called **Query Repeaterbook** should pop up. This window allows you to set the parameters of your query to RepeaterBook's database. The window doesn't contain as many filter options as the online database, but it still allows you to pull robust data sets. Here are the filter fields:
   - The **Country** field defaults to United States, but if you're located elsewhere, you can choose from more than 100 countries or polities.
   - **Service** allows you to find select repeaters for **Amateur** or **GMRS.**
   - **State/Province** allows you to select from all 50 states as well as the additional US territories.
   - You can specify **Latitude**, **Longitude,** and the **Distance** (in kilometers) from the provided coordinates.
   - The **Filter** box allows you to input a text string that RepeaterBook will match. It can be anything but will be most useful if it's easy for RepeaterBook to match, such as specifying a county name or emergency service affiliation. Trying to specify other location information, such as expressway proximity doesn't work as well.
   - **Limit Bands** – Allows you to specify which band(s) to pull data for. The options are:
     - **10 Meter Band**
     - **Meter Band**

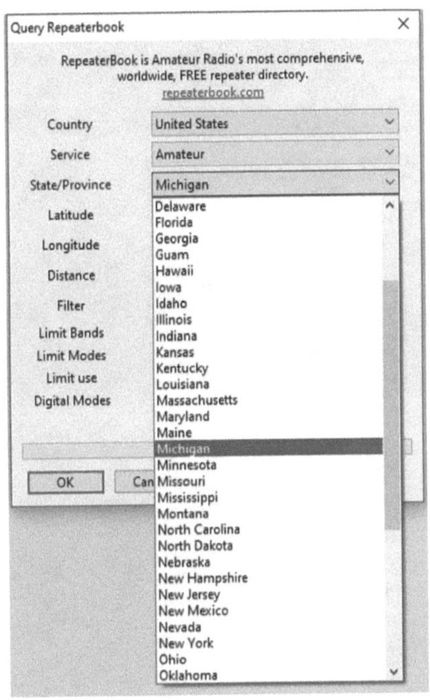

- 2 Meter Band
- 1.25 Meter Band
- 70 Centimeter Band
- 33 Centimeter Band
- 23 Centimeter Band
- 13 Centimeter Band

- **Limit Modes** – Allows you to specify the following:
  - **FM – Frequency modulation**
  - **DV – Digital voice**
  - **DMR – Digital Mobile Radio**
  - **DN – Yaesu System Fusion®**
- **Limit use** – Excludes private and closed repeaters from the list.
- **Digital Modes** – Shows dual mode (FM and digital) repeaters as FM in the data set.

3. Select whichever parameters make the most sense for you and click **OK** to execute the search.
4. A new file opens as a spreadsheet.

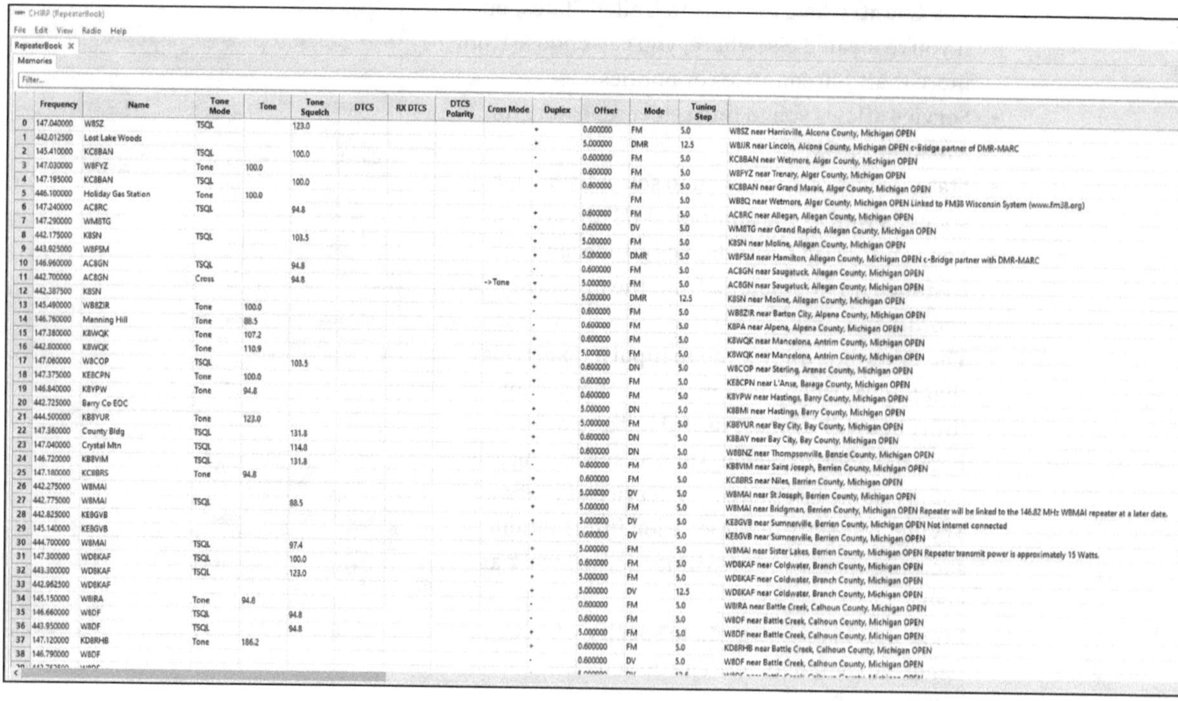

5. Press **Ctrl** and **A** at the same time to select all the data on the spreadsheet. Then right click and select **Copy** or press **Ctrl** and **C** at the same time.
6. You should have the last version of your settings and memories spreadsheet open, which should be v4. Click on the tab for your v4 spreadsheet near the top of your screen. If you don't have the file open, click on **File**, then select **Open...** and double click on your v4 file to open it.
7. Click on cell **6** in the **Frequency** column. Press **Ctrl** and **V** at the same time to paste the entries into your v4 spreadsheet, or you can right click on cell **6** and select **Paste.**
8. Press **Alt**, **Ctrl**, and **S** at the same time or select **Save as...** from the **File** dropdown to increment the version to v5.

Now, you have up to 128 repeaters ready to add to your radio! If you want to, you can standardize the **Name** fields and adjust the **Power** levels as necessary. Then just plug in your radio as you've done a few times and upload the memories. You're ready to get on the air with your Baofeng UV-82, thanks to *CHIRP* and RepeaterBook.

7.4.7 Programming Considerations

I have prepared some practical tips help you program your radio more effectively. *CHIRP* software helps you save time, label channels for easy navigation, and double-check tone settings to avoid communication issues.

When reprogramming your radio, you should think through what needs to be programmed and make a list of frequencies, tones, offsets, repeater-specific considerations, and regional band plans or regulations. The *ARRL Repeater Directory®*, RepeaterBook, or your local frequency coordinator, which you can find in the *ARRL Repeater Directory* are all good sources of information.

I find it useful to have a folder for each of my radios and a subfolder for local and travel files. Over time you may realize that you have many files to manage and managing them is easier if you start off with a solid system. For an example system, see item 6. Create a Logical Filing System in the following list.

1. PREPARE YOUR RADIO

Ensure *CHIRP* supports your specific Baofeng model (theUV-5R and UV-82 are both supported). Fully charge your radio to prevent interruptions during programming.

2. PROGRAMMING CABLE SETUP

Use a genuine programming cable or one with a reliable chipset. FTDI and Prolific chipsets are good bets. Always test the cable by plugging it into your computer before connecting the radio. Your computer should recognize the cable when you plug it in.

3. CHECK FOR SOFTWARE UPDATES REGULARLY

CHIRP's weekly updates add new radios to the compatibility list and provide bug fixes. It's worth keeping up with them, especially as your collection of radios grows and you start using the software in more specialized ways. Heck, as you use it, you may even think of new features that the software could use or run into unique problems. That's what the **Issues** log is for. If you have something to contribute, you can check out **https://chirpmyradio.com/projects/chirp/issues.**

4. FAMILIARIZE YOURSELF WITH THE USER INTERFACE

I've provided an in depth look at the various settings and memory options available in *CHIRP*, but even so, using the software and exploring it on your own will increase your proficiency with it. Spend some time exploring the dropdowns and tabs, pulling data, and programming your radio. As you do, you may discover different ways to use your radio.

5. BACK UP YOUR RADIO

Before making any changes to your radio configuration in *CHIRP*, click on **Radio** and select **Download from radio...** to retrieve and save a copy of your radio's current settings. Creating backups frequently is crucial in case you accidentally overwrite important settings.

6. CREATE A LOGICAL FILING SYSTEM

In addition, you should use a naming convention that has a date and version number to it. Creating a filing system with meaningful metadata will make finding the information you want much easier in the future when you've had time to forget exactly what you were thinking when you created the file.

I like to implement version numbers at the end of my file names, so that it's easy to tell which is most recent. I also like to use dates in my file names, so that I have a second grouping level. On any given day, I might make several different versions of a file with different memory groups. When mistakes happen, as they inevitably will, having that second level of grouping, allows me to quickly move through my files and narrow the problem by group. When I've found a group that works, I can quickly isolate what day the mistake occurred on and I can examine each file to figure out the exact location of the mistake.

I also have several radios that I program using *CHIRP*. I keep parent folders for each and subfolders based around use cases. That way I can quickly find files for my go-bag radios without having to wade through the files I use for travel to Connecticut or California. The system I'm outlining here is just an example for you. As you build out different use cases, program different radios, and spin through versions, you'll fine tune the system to meet your needs. What I'd like to encourage is developing a bit more granularity than dumping all of your *CHIRP* files into a single folder with nothing but version numbers to differentiate one file from another.

7. USEFUL *CHIRP* FEATURES

As I've demonstrated in earlier sections, **Copy** and **Paste** allow you to easily duplicate frequencies between channels and across spreadsheets. Using these features will increase the accuracy of your spreadsheets over manually inputting information.

The **Tone Mode** and **Tone** columns allow you to set up CTCSS tones for repeaters.

Add notes to each channel (e.g., Sun Valley ARES CH 1 or Newington ARC) in the comments tab. These don't display on the radio, but they can help you keep track of repeaters and other frequencies. For example, if your local Emergency Operations Center (EOC) operates on a set of frequencies, you could label them in more detail using this column if you volunteer or train with them.

You can also set the power level for specific frequencies using the **Power** column so that you don't have to mess around with your radio's settings when you're out in the field.

8. COMMON PROGRAMMING MISTAKES

Double check the settings in the **Duplex** column of the **Memories** spreadsheet to ensure the offset and direction (+/-) are correct for a given repeater. You can check the info in the spreadsheet against the information in RepeaterBook or the *ARRL Repeater Directory* or other sources to confirm it's correct.

Overwriting a memory slot can lose a previously saved frequency. This is a particular problem because you can drag lines up and down the spreadsheet with your mouse. While this feature can be handy, it does occasionally result in overwriting a frequency with a duplicate. Be careful using this feature when all 128 memory slots are full.

Ensure you're programming frequencies within the radio's supported range. RepeaterBook will let you pull a wide variety of bands. The radios in this book only support 2 meter and 70 centimeter operation.

The most common problem when trying to operate a repeater is that the tone is set incorrectly. If it doesn't seem like anyone is hearing you on a repeater that you can hear, double check the repeater's tone and make sure you have it set up correctly. Most of the time, this will clear up the problem.

9. EXPORT AND SHARE

Export your configuration to a file by clicking on the **File** dropdown and selecting **Save As....** This allows you to easily share settings with others or restore your configuration later.

Use the import feature to load pre-existing channel lists or configurations for your specific radio model.

CHIRP can pull repeater information from online sources such as RepeaterBook or Radio Reference, so you can quickly load up-to-date frequencies for your area.

Export settings to a CSV file for easy editing or archiving using a program like Microsoft's *Excel*.

10. IMPORT FROM OTHER RADIOS

The software also allows seamless file transfers between supported radios, offering additional flexibility.

I was able to take the files I use for my Yaesu FT-818 and export them as a CSV file, edit them in Excel, and then open the file in *CHIRP* and do the cut and paste across to build my UV-82 file with minimal typing.

11. TROUBLESHOOTING

Utilize the Help Documentation using the built-in help menu by visiting the *CHIRP* website. In the **Documentation** tab browse the **FAQs** and **BaofengUV82** pages for common issues.

If *CHIRP* won't recognize your radio, try:

- Rebooting your computer.
- Verifying you're using a compatible USB programming cable and drivers.
- Checking the COM port assignments in your operating system.
- Using the **Download from radio...** feature to verify the connection.

12. SKIP FUNCTION USEFUL FOR SCANNING

If you are going to use the **SCAN** function you will want to set the **Skip** column for the frequency to **S** to avoids stopping on certain channels, for example, it's annoying to get stuck listening to the weather channel frequency when you're looking for activity on other channels or frequencies.

7.4.8 Memories

In addition to pulling memory channels from online sources, *CHIRP* offers some standard memory files for things like **US FRS** and **GMRS Channels**, **US NOAA Weather Alert**, and **US Calling Frequencies** (for example) that can be used to add memory channels to the radio. To find these, click on **File** and select **Open Stock Config.** But I will warn you, these are universal lists and are not specific to Baofeng radios. For example, the **US Aviation Frequencies** will not work on the UV-5R and UV-82.

MANAGING FREQUENCIES

The UV-82 supports 128 channels and data entry can be time consuming, but copying data or pulling it from online sources can speed up the process. Well, what about keeping track of all those channels? A week after programming your radio, or a month or a year, how are you supposed to remember what all those frequencies were for? Sure, you can use the **Name** field, but that's limited to just a few characters. However, when you need more information, *CHIRP* allows for free-form text in the **Comment** field. It's here that you can make notes about the physical location of repeaters, the public service entity associated with each frequency, and anything else that's useful to you.

In addition, here are a couple more tips for keeping frequencies organized:

- You can simplify your programming of large memories lists by importing them.
- Use **Import from file...** for pre-built frequency lists from online sources or that you have created on your own in a spreadsheet.
- Organize channels by groups (e.g., repeaters, simplex, weather, emergency frequencies, and SOTA/POTA) to keep your radio easy to navigate.
- Program the frequencies you use most in memory channels 1 to 10
- Program your local weather channel in memory slot 0.

7.4.9 Memory Navigation Differences from the UV-5R

While the Baofeng UV-82 and UV-5R share many similarities, their memory navigation does have some key differences.

BAOFENG UV-5R MEMORY NAVIGATION

The UV-5R allows you to switch between **VFO** mode (manual frequency input) and **MR** mode (memory recall) by pressing the "VFO/MR" button. In **MR** mode, the channels are displayed numerically. You can scroll through the channels using the ▲/▼ keys. Channels can only be navigated sequentially (using ▲/▼ buttons) unless you manually enter the channel number via the keypad. While in memory mode, you cannot view the specific frequency, only the channel number (unless the memory is tagged with a name or frequency display is enabled).

In dual-display mode, you can toggle between the two displayed frequencies or channels by pressing the "A/B" button.

BAOFENG UV-82 MEMORY NAVIGATION

The UV-82 is unique in that one of its features is a **dual PTT system**, meaning the upper button transmits on the upper display frequency and the lower button transmits on the lower. This can make memory navigation slightly faster for users switching between two channels or bands often.

There are situations where a contester or public service volunteer could find the dual PTT system to be an advantage. I have the VHF calling channel on A and the UHF calling channel on B so I could easily switch back and forth without needing to use the band button to switch bands. I could be accepting incoming UHF (VFO B) traffic during a car rally from checkpoint 23 and acknowledge it back to checkpoint 22 and I could then transmit it on the VHF (VFO A) band to checkpoint 24.

Unlike the UV-5R, you toggle between **VFO mode** and **MR mode** by turning on the radio while holding the **MENU** button. Some users may find this annoying, and I will admit, it takes a little extra time while you wait for the UV-82 to boot up, but the wait is not long, and once you get used to the action, it easily becomes second nature. Another difference between the UV-82 and the UV-5R is the improved usability of allowing direct channel number entry on the keypad in **MR mode.** This feature reduces the time required for scrolling through channels sequentially.

Channel-specific settings (like power level or wide/narrowband mode) can be directly adjusted while in **MR** mode, making it slightly more convenient for advanced users.

The larger display on the UV-82 makes memory names or frequencies easier to read.

Overall, the UV-82 offers a slight edge over the UV-5R, particularly due to its dual PTT system, direct channel number entry, and improved display. However, for basic use, the UV-5R is still a reasonable choice and is easier for beginners to navigate.

7.5 Mini Usability Review

The Baofeng UV-82 is a solid upgrade for existing UV-5R users, with several feature improvements. The first thing you'll notice is the size of the radio, which, in my experience, is one of its strongest points. The larger form factor makes it much more comfortable to hold, and the bigger keys are easier to operate.

Another notable improvement is the improved audio quality. The upgraded speaker delivers clear and loud sound, making it a reliable choice for a variety of environments. The dual-band display is also clear and easy to read, and the buttons provide a satisfying tactile response, enhancing the overall user experience.

The dual PTT function is a unique feature that allows quick access to transmit on either the upper or lower display frequency. This makes the UV-82 versatile and user-friendly, accommodating both analog use and dual-frequency monitoring with ease.

While the UV-82 introduces a refreshed design, its operation and performance remain largely similar to the UV-5R. It offers slightly more power, though the difference isn't particularly noticeable during regular use.

However, the radio is prone to interference, especially in densely populated urban areas. This is due to the wide range of frequencies it can receive. In less RF-congested environments, the UV-82 performs reliably without major issues.

7.6 Accessories

Several accessories can enhance the functionality of the UV-82. Extra batteries, better antennas, and speaker-microphones are among the most popular. If you have a UV-5R, you should know that antennas can be shared, but speaker mics and batteries cannot.

Extra Batteries

- **Higher Capacity Batteries** – Consider purchasing high-capacity batteries (like 3800mAh) for extended operating time, especially during outdoor activities. These are available from third party vendors.

Charger Options

- **Multi-Desktop Charger** – A dual slot or multi-charger can charge multiple batteries or radios simultaneously.

Antenna Upgrades

- **Better Antennas** – Upgrading to a longer antenna (like a Nagoya NA-771 or Diamond SRH77CA) can improve reception and transmission range.
- **SMA-Female to SMA-Male Adapters** – These can help connect various antennas to the UV-82.

Microphones

- **Earpiece with PTT** – A simple earpiece with a dual push-to-talk (PTT) button allows for discreet communication, ideal for noisy environments.
- **Speaker/Microphone** – Any speaker/mic you purchase should support the UV-82's unique dual PTT function.

Carrying Cases and Straps

- **Protective Cases** – A soft or hard case can protect your UV-82 from scratches and damage.

USB Programming Cables

- Essential for programming frequencies, settings, and other configurations using software like *CHIRP*. Baofeng makes the K-Plug Programming Cable.

Software for Programming

- ***CHIRP* Software** – Free software that allows you to easily manage frequencies and settings on your UV-82 through a computer.
- **Baofeng CPS Software** – Proprietary software available from Baofeng.
- **RT Systems** – BAO-5R programming software with KY4-USB programming cable.

External Speakers

- A small external speaker can improve audio quality and make it easier to hear incoming transmissions.

Mobile Mounts and Holders

- If using the UV-82 in a vehicle, a dash or cup holder mount can keep it secure and accessible.

Battery Eliminators

- These allow you to power the UV-82 from your vehicle's power supply, eliminating the need for extra batteries during long trips. These devices are only available from third party vendors.

Adapters

- **SMA Adapters** – SMAs are not the only plug type used in amateur radio, so as you experiment with different antennas, you may find that you need different sorts of plugs. to connect to the UV-82 you'll need a female SMA plug on one side, but you may need PL-259s or other types of connections on the other. It's always a good idea to have a variety of connectors and adapters available.

8 The Baofeng DR-1801UV: A Digital Mobile Radio

DMR Tiers

DMR has three operating tiers, set by the European Telecommunications Standards Institute (ETSI).

Tier I Tier I is a set of 16 channels between 446 and 446.2 MHz set aside for unlicensed use with a maximum power output of 500 mW.

Tier II This is the ham radio tier. Tier II radios can operate between 66 MHz and 960 MHz and can use repeaters to increase the communication's geographic range.

Tier III Covers the same frequencies as Tier II, but the radios operate in "trunked" mode, meaning that the frequency the radio uses is automatically assigned. This minimizes the bandwidth required for communication and the likelihood of interference.

8.1 Overview

The Baofeng DR-1801UV is a dual-band, dual-mode handheld radio that bridges the gap between analog and digital communication. Supporting both traditional FM analog and Digital Mobile Radio (DMR), it offers a versatile solution for amateur radio operators transitioning to digital modes. Its compact, lightweight design features a LCD screen, programmable keys, and a user-friendly interface, making it an accessible option for new and experienced radio enthusiasts alike.

The DMR radio has channel scan, manual programming, DTMF, VFO, SMS, emergency alarm, digital remote monitor mode, CTCSS, and DMR encryption functions. There are four customizable panel keys for faster access to those features.

DR-1801UV has high-performance energy saving and power saving. The 2200 mAh battery supports up to 17 hours of working time and 30 hours of standby time and can be fully charged in only 4 hours.

While it may lack some advanced features found in higher-end DMR radios, the DR-1801UV delivers reliable performance at an affordable price point. Notably, it is DMO True 2 Time Slot compliant, enabling dual-slot communication in direct mode. This feature allows two groups to communicate simultaneously on the same frequency, doubling efficiency and conserving frequency resources. The radio is also compatible with Motorola MOTOTRBO Tier I & II systems and other DMR radios, ensuring broad interoperability.

Batteries are specific to the DR-1801UV, but other peripherals such as external antennas and speaker mics are interchangeable with most Baofeng models, adding to its practicality and ease of use.

The contents of the Baofeng DR-1801UV box vary depending on the package, but typically include the following items:

8.1.1 Contents

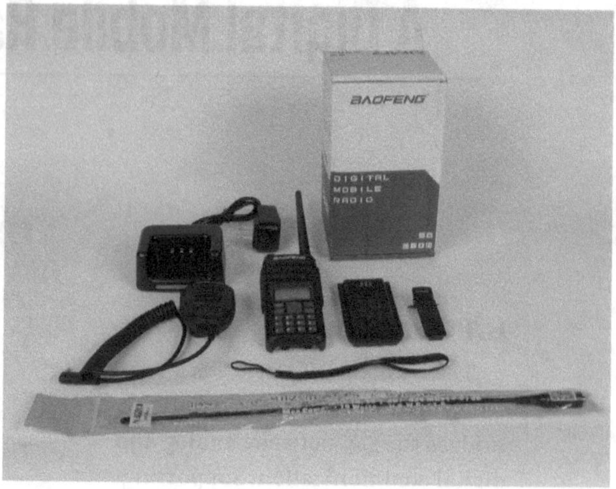

- **Baofeng DR-1801UV Radio Unit** – The handheld transceiver itself. My model features the A6 processor.
- **Antenna** – Usually a flexible 5" dual-band antenna (VHF/UHF) with an SMA-Female connector.
- **Battery Pack** – A rechargeable lithium-ion battery, typically rated at 22000 mAh.
- **Belt Clip** – A screw-on clip for attaching the radio to your belt or pocket.
- **Hand Strap** – A lanyard-style wrist strap for additional portability.
- **Desktop Charger** – A cradle-style drop-in charger with AC adapter for charging the battery.
- **Earphone/Microphone** – For private listening and has dual PTT buttons
- **User Manual** – A basic instruction manual that explains the radio's features, programming, and operation.

My radio purchase included a speaker/microphone and a Nagoya NA-771 gain antenna as a nice surprise in the box. However, I was expecting a programming cable and one was not supplied.

BASIC SPECIFICATIONS

- **Frequency Range:** 136 – 174 MHz (VHF), 400 – 480 MHz (UHF).
- **Power Output:** 1 W / 5 W selectable.
- **Channels:** 1,024 programmable channels.
- **Battery:** 2200 mAh Li-Ion rechargeable battery (7.4V).
- **Antenna:** Dual band SMA female.
- **Display:** Dual-band display.
- **Dimensions:** 5.2 × 2.4 × 1.4 inches.
- **Weight:** Approximately 9.6 ounces (including battery).
- **Connectivity:** 2-pin Kenwood-compatible audio jack
- **Additional Features:** SMS function and the advanced AMBE+2TM voice processing technology to achieve improved call quality.

8.1.2 Basic Operation and Getting Started with the DR-1801UV

Here's a comparison of the Baofeng DR-1801UV and Baofeng UV-5R in a tabular format, highlighting the key differences.

Table 8.1

Baofeng DR-1801UV and UV-5R Feature Comparison

Feature	Baofeng DR-1801UV	Baofeng UV-5R
Frequency Range	VHF: 136 – 174 MHz, UHF: 400 – 480 MHz	VHF: 136 – 174 MHz, UHF: 400 – 520 MHz
Modes Supported	FM analog, DMR Tier I and II	FM analog only
Display	Color LCD display	Monochrome LCD display
Dual PTT Function	Yes, dual PTT buttons, DMO	No, single PTT button
Battery	2200 mAh Li-ion rechargeable battery	1800 mAh Li-ion rechargeable battery
Power Output	5W (VHF), 5W (UHF)	4W (VHF), 4W (UHF)
Channel Capacity	1024 memory channels	128 memory channels
Programming	Keypad or PC (USB cable & software)	Keypad or PC (USB cable & software)
Audio Output	Clear, loud audio with enhanced speaker	Clear audio, but less powerful
Digital Mode Compatibility	Yes, DMR, also supports direct mode (DMO)	No, analog only
Size and Weight	5.3 x 2.3 x 1.3 inches, 0.6 lbs	4.3 x 2.3 x 1.3 inches, 0.3 lbs
Price Range	Typically higher due to DMR capabilities.	Lower, entry-level price for analog use
Additional Features	True 2 Time Slot for DMR, Direct Mode (DMO)	Basic analog functions, many accessories

You can use your UV-5R antenna and speaker/mic accessories with the DR-1801UV but they cannot share battery packs. The charge time is approximately 4 hours for the 2200 mAh DM-8 battery.

8.2 Button-by-Button Walkthrough

Here I'll cover each button's functionality and how to use it effectively. There are two programmable panel keys you can customize (side key 1 and side key 2).

8.2.1 Front Panel

The DR-1801UV has an informative two-line LCD display that shows frequency, battery level, signal strength, mode indicators, power level, and any activated features (like CTCSS tones). The display backlight can be set to different colors for various modes.

The Baofeng DR-1801UV features a straightforward front panel layout.

8.2.2 Top Panel

- **On/Off Volume Knob** – Combined with the selector knob. Turn clockwise to power on and adjust the volume.
- **Antenna Port** – Allows you to connect the included or a third-party antenna.
- **LED Indicator** – Has two settings:
  - Red – The radio is transmitting.
  - Green – The radio is receiving a signal.
- **SOS (Orange)** – Preprogrammed as an emergency alarm button.

8.2.3 Side Panel Buttons

1. **PTT (Push-to-Talk) Button** – Large button on the left side. Press to transmit, release to return to receive mode.
2. **Side Key 1 (Top Button — up arrow)** – Customizable function, often pre-programmed for features like Monitor (open squelch) or Flashlight activation.
3. **Side Key 2 (Bottom Button — down arrow)** – Another customizable key, often assigned to Alarm or other user-configured functions. I like to have Monitor and the High/LOW power functions assigned to these keys.
4. **Speaker/Microphone Jack** – Two-pin Kenwood-style jack for external accessories like a programming cable, speaker mic, or headset. Using the *Baofeng Customer Programming Software* (CPS), specific keys can be assigned for emergency alarm, high or low power, monitor, one-touch access, and repeater talk around.

8.2.4 Display

The DR-1801-UV has an informative two-line LCD display that shows key information such as the frequency, battery level, signal strength, mode indicators, power level, and any activated features (like CTCSS tones). The display backlight can be set to different colors for various modes.

Examining the LCD screen from the upper left corner and moving right line by line, the antenna in the upper left indicates the signal strength just like a cell phone, with more lines indicating a stronger signal. Beside that the H indicates that the radio is set to transmit at high power, and the square waves beside it mean that the radio is in digital mode. Analog mode is indicated by a single sine wave. The plus indicates that the offset is set to positive rather than negative, and the battery indicator is full when displaying three lines. The arrow on the left hand side of the second line means that VFO A is selected on the frequency 442.10000. The 1 on the far right indicates the channel. The bottom line shows the frequency VFO B is set to and the channel.

MENU BUTTON

Accesses the radio's main settings and configuration menu. A short press opens the menu system where you can change settings like frequencies, power level, squelch, tones, etc. It also acts as a confirm or OK button when selecting options in the menu. You can use the 4-way directional buttons to navigate the menu options.

VFO/MR BUTTON (FREQUENCY MODE/MEMORY RECALL)

Toggles between Frequency Mode (VFO) and Channel Mode (MR).

- **VFO Mode** – Allows manual entry of frequencies or scanning through the band.
- **Memory Mode (MR)** – Lets you switch between pre-programmed channels stored in memory.

BACK BUTTON (EXIT)

If you're navigating a menu or configuring settings, pressing **BACK** will return you to the previous screen or main display. It will also cancel an entry.

A/B BUTTON

Switches between the two operating bands (or channels) displayed on the screen. The radio supports dual standby, so you can monitor two frequencies or channels simultaneously.

A Band: The top line on the display (primary).

B Band: The bottom line on the display (secondary).

4-WAY DIRECTIONAL BUTTON (▲/▼/◀/▶)

Multifunction navigation tool for menu control, frequency/channel changes, and quick settings.

▲ – Scrolls up through menu options, increases frequencies, or selects the next channel.

▼ – Scrolls down through menu options, decreases frequencies, or selects the previous channel.

◀ – Often used for decreasing numerical values in settings or toggling specific options.

▶ – Often used for increasing numerical values in settings or toggling specific options.

The A/B and VFO/MR Buttons

The A/B button ensures you're working on the correct frequency or channel while the VFO/MR button lets you quickly switch between memory and frequency modes.

HOW THEY WORK TOGETHER

The **MENU**, **VFO/MR**, and **BACK** buttons work in tandem with the 4-way directional button for smooth operation.

- Press **MENU** to open the settings.
- Use the ▲/▼/◄/► buttons to navigate and adjust settings.
- Press **MENU** again to confirm changes or **Back** to cancel and exit.

8.3 Programming Differences between the UV-5R and the DR-1801UV

The DR-1801UV and UV-5R share some similarities in basic functionality but programming them differs significantly due to the digital DMR capabilities of the DR-1801UV versus the analog-only UV-5R.

Due to the complexity of DMR programming, you must use the Baofeng CPS. The CPS includes additional sections for managing digital contacts, talk groups (TG), and DMR IDs. Also, each memory channel must be defined as either analog or digital.

Below is a breakdown of the key programming differences between the two radios.

Table 8.2

Programming Differences Between the UV-5R and The DR-1801UV

Feature	UV-5R	DR-1801UV
Programming Software	CHIRP or Baofeng CPS	Baofeng DMR CPS
Digital Parameters	N/A	Color Code, Time Slot, Talk Group
Manual Programming	Simple	More complex (digital settings)
Memory Structure	Analog channels only	Analog and Digital modes
Scanning	Analog only	Analog & Digital (Talk Group-specific)
Power Output	5W (VHF), 5W (UHF)	4W (VHF), 4W (UHF)

8.3.1 Baofeng DR-1801UV Memory Navigation

Like the UV-5R, you can toggle between VFO mode and MR mode using the **VFO/MR** button. The DR-1801UV improves usability by allowing direct entry of the channel number on the keypad in MR mode, reducing the need to scroll through channels sequentially.

The DR-1801UV supports software-driven memory banks via tools like the *Baofeng DR-1801_CPS*, enabling users to organize and recall channels by bank. This is particularly useful if you frequently switch between different sets of channels.

Channel-specific settings (like power level or wide/narrowband mode) can be directly adjusted while in MR mode, making it slightly more convenient for advanced users.

The larger display on the DR-1801UV makes memory names or frequencies easier to read.

8.3.2 Differences between the UV-5R and the DR-1801UV in Memory Navigation

The memory navigation differences between the DR-1801UV and the Baofeng UV-5R primarily stem from the DR-1801UV's ability to handle both digital (DMR) and analog channels, while the UV-5R is an analog-only radio.

All channels are stored in a single list on the UV-5R. Channels are navigated sequentially using the UP/DOWN arrow keys or entered directly using the keypad.

The 1801's memory channels can be analog or digital (DMR). Digital channels include additional settings like color codes, time slots, and talk groups. Navigation involves distinguishing between digital contacts (DMR) and analog frequencies, which adds complexity. Channels are organized into Zones.

The UV-5R will display the channel name or frequency whereas the 1801 can display additional digital-specific information, such as talk groups (TG), contacts, zones, and color codes.

Front panel programming is easily accomplished on the UV-5R but DMR requires the CPS to be used, as front panel programming is very complex.

8.4 An Overview of Digital Handheld Modes

Digital radio modes offer advanced features, improved audio clarity, and networked communication capabilities compared to traditional analog systems. These modes are increasingly popular among amateur radio operators and include DMR, Yaesu System Fusion (YSF), D-STAR®, NXDN, and many more. Each mode has unique features and operational requirements, providing a range of options for different use cases.

Each of these networks has its strengths—DMR-MARC for structured repeater-based communication, BrandMeister for flexibility and feature-rich integration, and TGIF for its community-driven and customizable approach. DMR-MARC shares common talk groups with BrandMeister like TAC 310, 311, 312, State Groups and Nationwide 3100.

Choosing a digital mode is always a difficult decision but is usually defined by the radio purchase. Icom users will benefit from D-STAR and Yaesu users use Fusion to communicate. DMR is supported by Motorola, Hytera and many Chinese brands. Baofeng offers a stable, robust and DMR-compliant radio at a cost-effective price.

The biggest advantage of the digital radio modes is that with their internet connected repeater platforms they have an ability for worldwide networking. This allows one to expand their communication range and be able to have conversations with hams around the world through your handheld. No longer is a 100 W HF radio and dipole antenna a requirement, when a Baofeng DMR-enabled radio can help you contact DX (long distance) stations.

For example, I had a wonderful conversation on the Worldwide Talk Group with a ham in Austria while sitting in my car in the car wash line up in Ontario, Canada. The communication was crystal clear, and my radio displayed his call sign while we talked. My local DMR repeater is not too far away, so the stock antenna on the 1801 worked as expected. For longer range I use a mobile antenna to access the more distant DMR repeaters downtown.

Here's a brief explanation of a few major DMR networks available.

8.4.1 DMR-MARC (Motorola Amateur Radio Club)

The DMR-MARC (Motorola Amateur Radio Club) network is one of the original DMR networks, built on Motorola's DMR technology. It focuses on reliability and a more structured approach to managing repeaters and users. DMR-MARC connects amateur radio operators worldwide through a linked repeater network, offering a variety of regional, national, and worldwide talk groups. Unlike some other networks, DMR-MARC is more centralized and typically requires authorization to add repeaters or users to its system. Its emphasis is on professional-grade operation, making it popular with hams who value structure and consistency.

8.4.2 BrandMeister

This is a more flexible and open DMR network, designed to give users more control over their connections. It allows operators to dynamically link repeaters and hotspots to hundreds of talk groups worldwide, including regional, national, and special-interest groups. With a focus on user-driven features, BrandMeister provides tools like a self-service dashboard, the ability to patch talk groups to other digital modes (such as D-STAR and YSF), and integration with personal hotspots. Its decentralized nature and wide compatibility with hotspots make it one of the most popular networks for amateur DMR users.

8.4.3 TGIF Network

The TGIF Network is a newer and more community-focused DMR network that is hotspot-friendly and offers customizable features. It caters to special-interest groups, allowing users to create custom talk groups that don't need to follow the rigid structures of larger networks. TGIF is particularly known for its friendliness to experimental setups and personal use, with tools to manage private and public talk groups easily.

8.4.4 DMR (Digital Mobile Radio)

DMR is a time-division multiple access (TDMA)-based system, allowing two simultaneous conversations on a single frequency. It is popular for its cost-effectiveness and global connectivity through networks like BrandMeister. DMR supports talk groups, private calls, and text messaging, making it a versatile mode for digital communication.

8.4.5 Yaesu System Fusion (YSF)

YSF Fusion is a proprietary mode developed by Yaesu, offering a seamless blend of analog and digital communication. It features an easy-to-use interface and supports group calls, text messaging, and image sharing. YSF is compatible with WIRES-X, enabling internet linking and extended communication reach.

8.4.6 Digital Smart Technologies for Amateur Radio (D-STAR®)

D-STAR, developed by the Japan Amateur Radio League (JARL), is a pioneer in digital radio modes. It provides low-bandwidth digital voice, data, and GPS capabilities, facilitating efficient communication. D-STAR's reflector system enables worldwide communication by linking repeaters through the internet. Icom and Kenwood offer D-STAR enabled radios.

8.4.7 Next-Generation Digital Narrowband (NXDN)

NXDN is a narrowband digital mode operating at 6.25 kHz channel spacing, offering excellent spectrum efficiency. Initially designed for commercial and public safety use, it has found its place in ham radio. This digital radio protocol is Kenwood branded. Icom's Digital Advanced System, (IDAS) is Icom's digital Land Mobile Radio platform with a scalable architecture. At the core of IDAS is the NXDN protocol.

8.5 An Introduction to DMR

8.5.1 Step-by-Step Guide to Obtaining a DMR ID

DMR requires user registration and an assignment of an ID for DMR operations. This ID is passed by the radio for access to the various DMR trust servers. It is also used for the user database that allows radio call signs to be displayed during conversation. You will need this DMR ID when you program your radio.

Obtaining a DMR ID is a straightforward process, but it does require registration and verification through the official Radio ID organization.

Here's how you can get your DMR ID:

1. Go to: **https://www.radioid.net/**. It is the primary authority for DMR IDs which manages the assignment of DMR IDs globally.
2. Register for a new DMR ID.
   - Click on **Register** in the navigation menu.

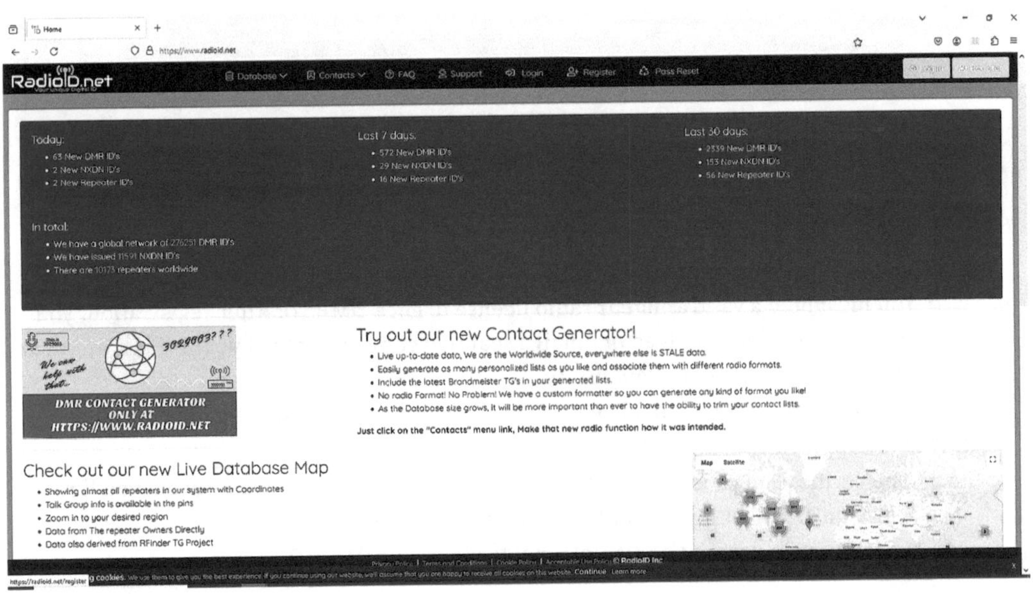

- Fill out the registration form with the following information:
    - Callsign
    - EMail
    - Password
    - Country/Territory
- Agree to the terms and conditions, then submit the form.

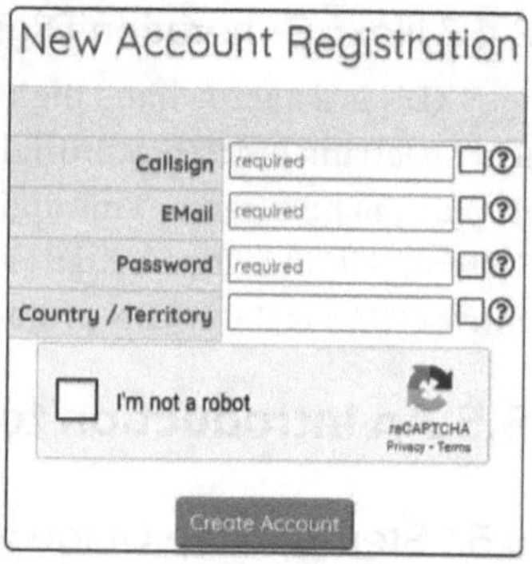

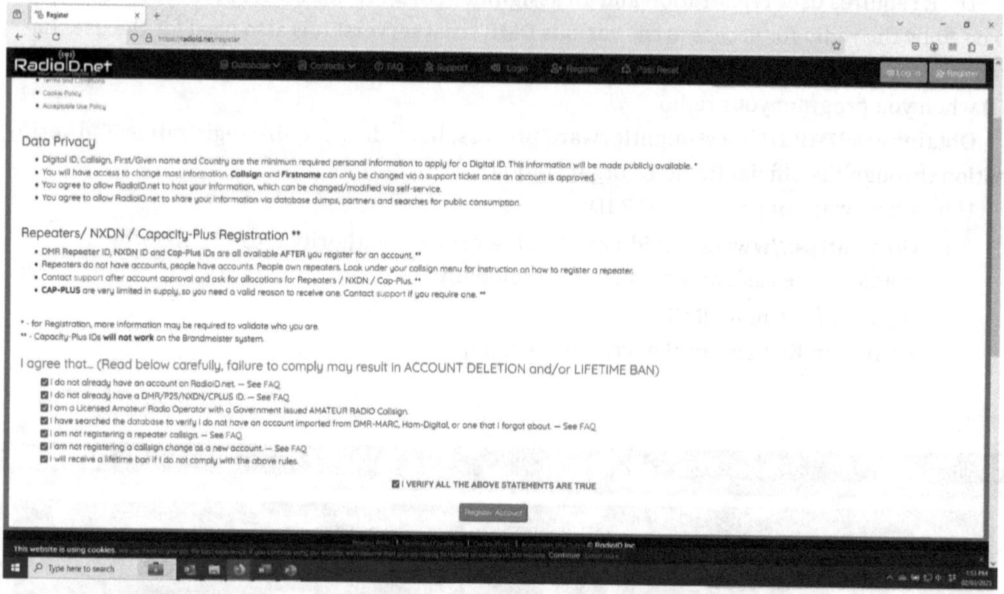

3. Submit proof of your amateur radio license.
    - You must hold a valid amateur radio license to get a DMR ID. After registration, you may be required to upload or email a scanned copy or photo of your official amateur radio license as proof of eligibility.
    - This step ensures that only licensed operators are issued DMR IDs.
4. Wait for approval.
    - The verification process usually takes two to three days. Once approved, you'll receive an email with your DMR ID and account information.
    - Log in to your account on RadioID.net to view your DMR ID.

8.5.2 DMR Code Plugs

Programming a DMR radio is more complex than an analog radio due to the additional information contained in the digital radio protocol. You'll need to program time slots, color codes, and talk groups, all of which are different from the information you need to program an analog radio. But as I walk you through programming a DMR radio the whole process will start to make more sense.

First, let's talk about code plugs. A DMR code plug is a data file that contains all the programming settings for a digital mobile radio. It serves as the radio's configuration blueprint, enabling seamless operation on specific networks or repeaters by organizing all the necessary information. In general, code plugs simplify radio use and ensure consistency and efficiency by preloading all the required information for communication. Because the 1801UV supports analog communications modes, the code plug for the radio can also include FM repeater or simplex configurations. You can build custom code plugs to meet your individual needs, but if you'd like, a variety of ham radio brands, local clubs, and individuals provide lists of DMR IDs for download. To locate these, simply search online for something like, "download Baofeng DMR contact list."

To create or edit code plugs, you need specific programming software that's usually unique to a brand or radio. The *Baofeng DR-1801 Programming Software* is what you'll need to program the radio in this book. The next section will detail installing the programming software.

Contents of a DMR Code Plug

A code plug is broken down into units I like to call *blocks,* a single block contains a single type of data. A typical code plug includes five programming blocks that need to be programmed in the following sequence:

1. CONTACT LIST

The contact list in the *DR1801 Programming Software* is just what it sounds like, a list of people or groups that you preprogram into your radio. A contact can be either an individual (referred to as a **Private Call**), a group of people (**Group Call**), or to anyone on a timeslot (**All Call** see the sidebar Time Slots in DMR). If you know someone's DMR ID, you can program a **Private Call** to that person. This is like calling for a specific call sign on a repeater or over HF instead of calling CQ

Time Slots in DMR

Time slots split up channels in DMR. Basically, sound information is converted into a digital packet and sent in one of two time slots over the internet. The time slots are referred to as TS1 and TS2. Splitting things up this way allows for multiple calls on the same channel. Anyone in the same time slot and in a given talk group (TG) will hear a Group Call. People in different time slots or TGs won't. The All Call group goes out not just to those in the same TG, but to all those in the same time slot, regardless of TG. It is not a two-way call, but instead a broadcast out to the entire time slot. In ham radio, All Call is not used, though it does have use in certain commercial scenarios. Think of the call for "Clean up on aisle four," in a grocery store. The people out gathering carts in the parking lot don't need to hear it, but everyone working the floor in the store should hear it.

(calling all stations). To find specific stations, you can search the database on **https://radioid.net.** As of this writing, the database contained more than 276,000 DMR IDs. One of the limitations of the DR-1801UV is that Baofeng's programming software doesn't allow you to download the database or even sections of it. More advanced DMR radios can store enormous numbers of contacts, allowing you to see the call signs of anyone you talk to via DMR. For the DR-1801UV, you must program each contact manually, which is great for learning about what goes into a DMR contact and talking around town, but is impractical to do for the worldwide DMR contact list.

2. TALK GROUPS

Talk groups (TG) are basically channels or rooms (different DMR systems use different terms to explain things) that you can join when you connect to the network. For example, TG 91 is the worldwide talk group and the North American talk group is TG 3. Not all repeaters carry all talk groups and the repeater owner will publish information about which talk groups you can access through the repeater. The ARRL HQ DMR repeater that I'll use as an example allows users to access the following talk groups:

- Time Slot #1 – World Wide (PTT on demand 5 min inactivity timeout)
- Time Slot #1 – World Wide English
- Time Slot #1 – North America
- Time Slot #1 – Northeast
- Time Slot #1 – TAC310 PTT on demand 5 minute inactivity time
- Time Slot #1 – Parrot audio test

Time Slot 2 on this repeater is dedicated to ARES communications and gives the user access to Connecticut ARES-affiliated repeaters in specific parts of the state. The parrot audio test allows you to record and play back a brief message so you can check your audio.

Full Time talk groups are always available for monitoring (like the World Wide or World Wide English). Others may crop up at specified times or randomly, depending on the members and its purpose. PTT TGs require you to specify the amount of time after you push the PTT button that the TG will remain active, as in the first and fifth bullets above.

3. CHANNELS

Channels create the list that you scroll through on your radio. They are similar to memory channels that can store repeater configuration information, a name, etc. You can specify each channel as digital or analog. You will need to specify the following information for each channel:

- Receive and transmit frequencies for repeaters or simplex use.
- Offset or repeater shift details.
- Color Codes (0 – 15) which are similar to CTCSS in analog systems.
- Transmit Criteria set to Channel Free, Color Code Always.
- Time Slots (TS1 and TS2) the code plug determines which slot to use for each channel.

4. ZONES

Zones allow you to group channels, making the menus easier to navigate. If you have multiple hotspots or repeaters that you intend to use, for example, if you've got a hotspot at your house, in your truck, and at the shop, you might group those in a single zone. But if there's a DMR repeater you use from home, your local hotspot with perhaps some different TGs enabled, and then an analog repeater you like to use to hop on the nightly ARES net, then they all might go together in a single zone labeled "Home," while the truck and shop might be in their own zone. Really, zones are about what makes the most sense to you and makes your radio the easiest to navigate. If you only have a couple of channels, then maybe you only have one zone. Zone names can be up to 16 characters long.

5. SCAN LISTS

Scan lists are exactly what they sound like, too. They contain lists of channels to scan when you enable the scan function. This feature is a little tricky because you have to assign each *channel* to a scan list. So, when you're modifying or building your channel list, that's when you need to set which scan list the channel is a part of.

8.5.3 Information You'll Need before Programming Your DMR Radio

Gather Information about the Repeater

Before programming, you'll need the following details about the DMR repeater:

1. Repeater Frequency:
   - **RX (Receive) Frequency:** Frequency on which the repeater transmits.
   - **TX (Transmit) Frequency**: Frequency on which your radio transmits.
2. **Color Code** – DMR repeaters use a color code (0 – 15), similar to CTCSS in analog systems.
3. **Time Slot** – DMR repeaters have two time slots:
   - **Time Slot 1 (TS1)**: Often used for wide-area or networked talk groups.
   - **Time Slot 2 (TS2)**: Often used for local or regional talk groups.
4. **Talk Group** – Identify the talk group(s) you wish to access (e.g., TG91 for Worldwide, TG3148 for Texas).
5. **Repeater Offset** – The difference between RX and TX frequencies (e.g., ±5 MHz for 70cm or ±600 kHz for 2m).
6. **Call Types** – DMR supports three types of calls (for more information, see the previous section on contact lists and the sidebar Time Slots in DMR):
   - **Private Call:** Between two users.
   - **Group Call:** Communication within a Talk Group. This is the standard setting.
   - **All Call:** Broadcast to all users on the same timeslot.

As an example, I will walk you through programming the W1HQ DMR repeater located at ARRL Headquarters. I should also note, that unless you are within a few miles of Newington, CT, where the W1HQ repeater is located, you won't be able to test your radio once you've programmed it with the settings in the example that follows. If you haven't yet, you should find a local DMR repeater within range of your handheld and program your radio with its information either alongside or in place of the W1HQ repeater. That way, you'll be able to verify your talk groups and time slots are correct when you've finished programming your radio.

8.5.4 *Baofeng DR-1801 Programming Software* Installation

As mentioned earlier, to program the 1801UV, you'll need to install and use the *Baofeng DR-1801 Programming Software* on a Windows computer.

Here are the basics of programming your radio and some tips to enhance your DMR experience.

STEP 1: DOWNLOAD THE *BAOFENG DR-1801 PROGRAMMING SOFTWARE*

1. Go to the official Baofeng website at **www.baofengradio.com/.**
2. Hover over the **Support** tab and click on **Download.**
3. Locate the **DM Series** row.
4. Click on the DR-1801UV button and download the manual and software from the popup.

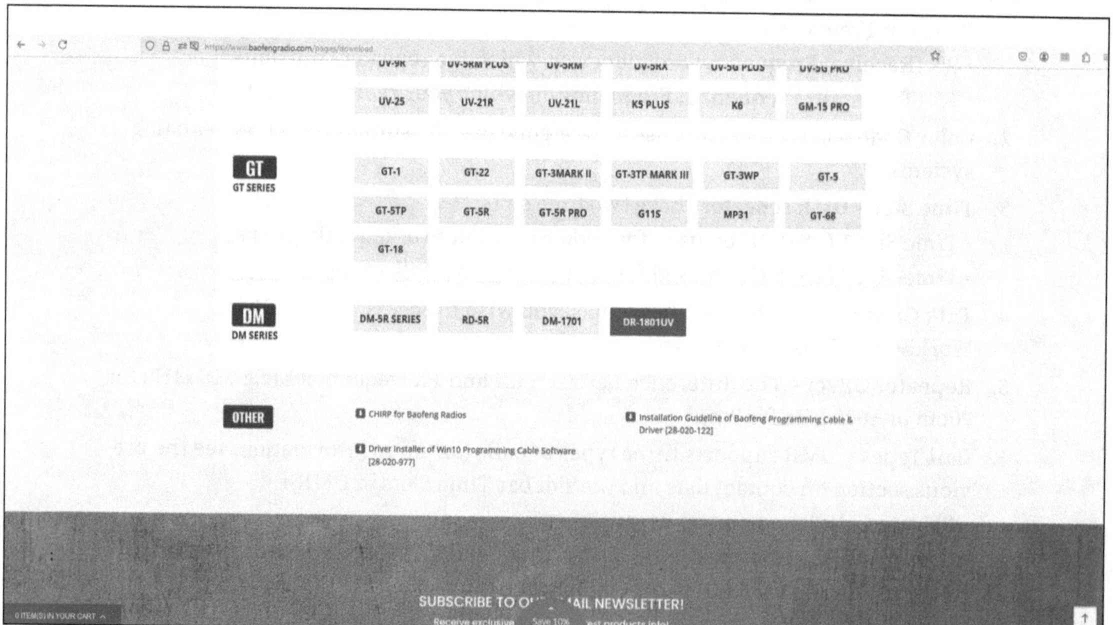

STEP 2: INSTALL THE *BAOFENG DR-1801 PROGRAMMING SOFTWARE* ON A WINDOWS COMPUTER

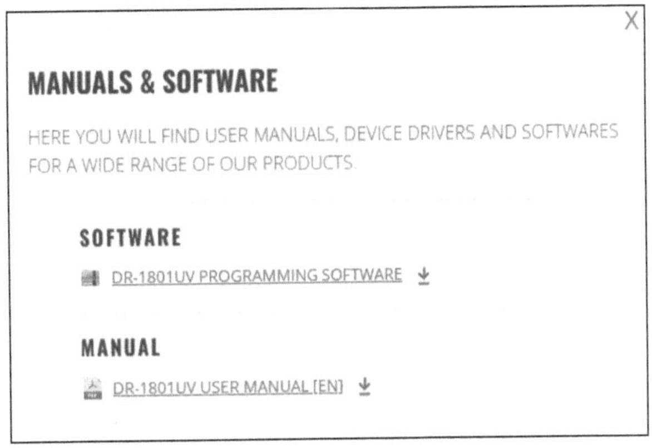

1. Once downloaded extract the compressed files. Locate the installer file, called DR-1801UV_CPS, in your **Downloads** folder and double-click it to start the installation.
2. Follow the directions provided in the **Installation Wizard.**
3. Once the installation is complete, click **Finish** to exit the installer.
4. You will also need to install the Microsft.net framework. This should be included in the zipped file along with the *Baofeng DR-1801UV_CPS.*
5. You may also need to download the USB driver for the programming cable. If you need directions for installing drivers for the programming cable, see section 7.4.3, Step 3.

STEP 3: LAUNCH THE *BAOFENG DR-1801 PROGRAMMING SOFTWARE*

1. Locate the *Baofeng DR-1801UV_CPS* icon on your desktop and double click it to launch the program.
2. Connect your radio to the PC using a Baofeng official USB programming cable.

Rotate the volume control to halfway and ensure you are not on a channel that has traffic. If your radio is receiving transmissions, the audio will cause programming errors.

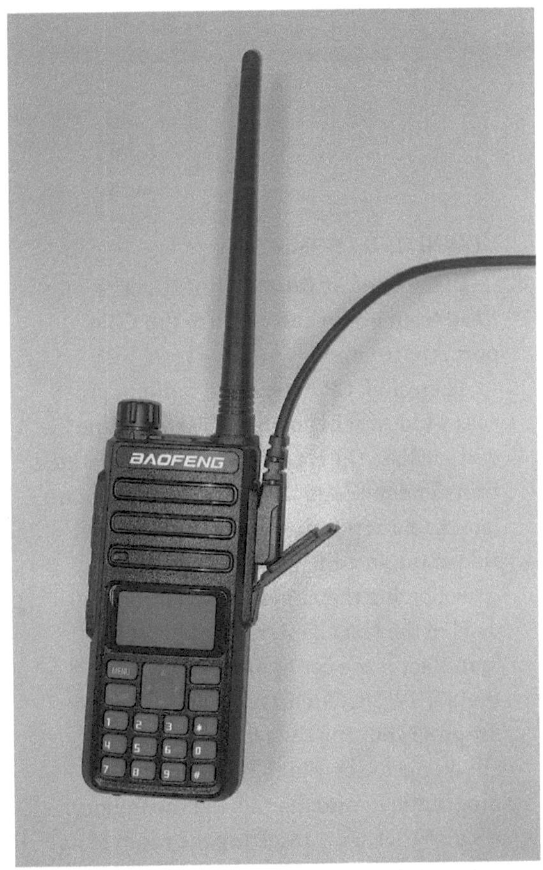

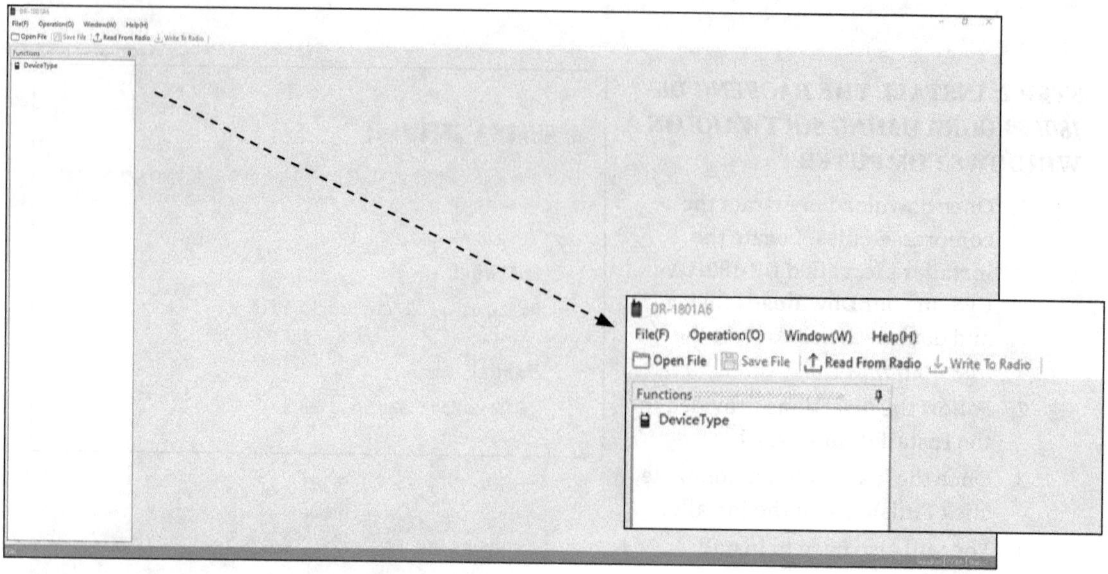

DOWNLOAD FROM RADIO

In the *Baofeng DR-1801UV_CPS*, click on **Read from Radio** and confirm the COM port is correct.

Then click **OK** to read the radio. A process window will pop up. The status line will indicate the transfer status. When the transfer completes, a list of tabs will populate the **Functions** panel on the left-hand side of the screen.

Following the naming convention I used in the last chapter, I'll name this manufacturer's configuration file VE3IPS_DR1801UV_20250106_OEM. This file isn't for programming the radio, but rather it allows me to see which frequencies are programmed into the radio at the factory. It's also useful for resetting the radio to its default settings.

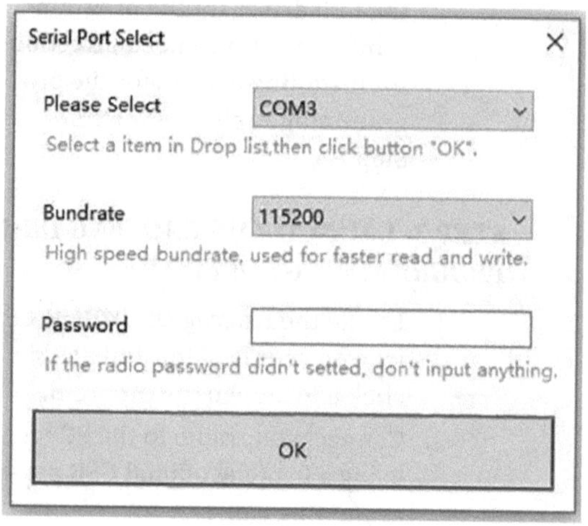

8.5.5 Modify the Basic Radio Settings

The **Basic Configs** tab in the **Functions** panel is where you go to input the radio's key configuration and user interface settings. To get started, let's go through a few must-input items in **Basic Configs.**

1. In the **Functions** list in the left-hand pane, click on Basic Configs.
   - The **Basic Configs** tab should open.
2. In the **Radio Name** field, enter your call sign.
3. In the **Radio Id** field, enter your seven-digit DMR Radio ID.
4. Configure other basic settings such as power level, squelch, timeout timer (TOTs), BLCO etc.

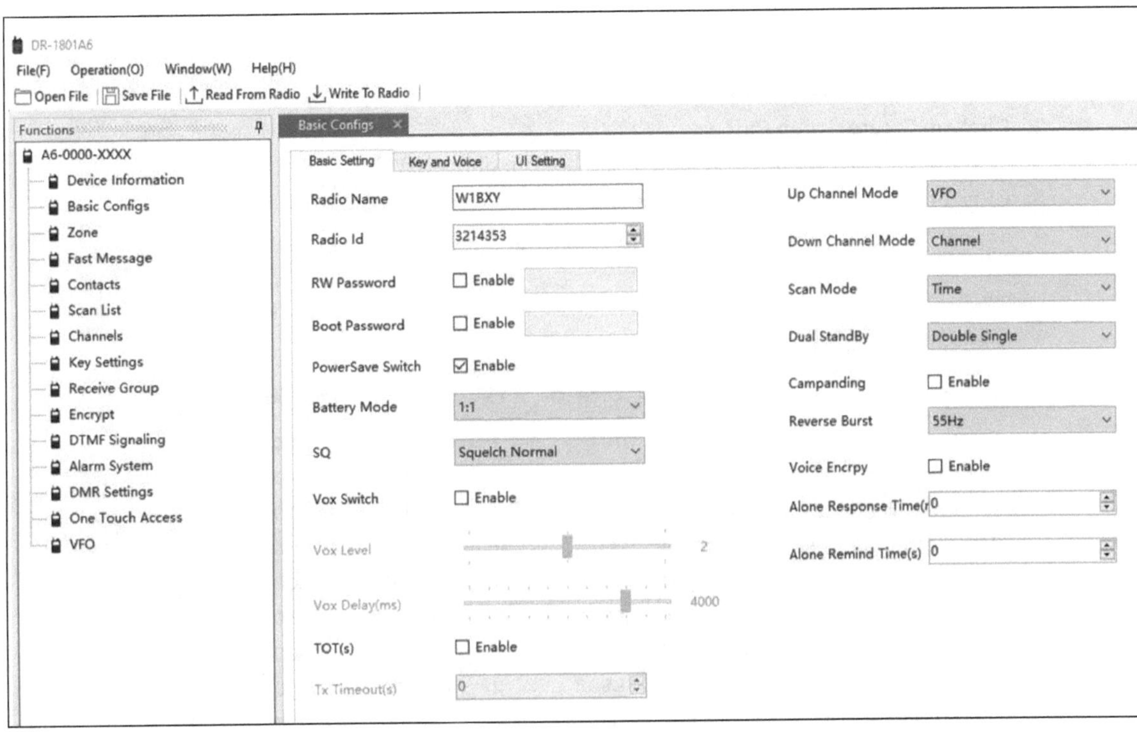

5. Click on the **File** dropdown and select **Save As...** so you can save these settings using your naming convention. Mine is VE3IPS_DR1801UV_20250106_DMR_v1. This is the file to start with for later programming.

8.5.6 Create a Contact List

The contact is the base unit or building block of DMR programming. It contains the basic information you need to connect to talk groups or individuals. Each contact contains the **Name,**

Type, and **Number** of the talk group you want to connect to. The contact name can be anything you want, but should probably reflect who you can reach on it. The type is almost always going to be **Group Call.** The number is the DMR ID assigned to the talk group. Repeaters and hotspots only connect to certain talk groups, so before you start programming talk groups into your radio, it makes sense to review the repeaters and hotspots you will be connecting to and aligning your available talk groups with theirs. In addition to the available talk groups, you will need to know which time slot the TG has been assigned to on the repeater. You will need to program this information into the contact when you create it. To line up this information before you start programming, you should consult **www.repeaterbook.com** or ask your local frequency coordinator.

Let's take a look at the W1HQ DMR repeater. I obtained the following information from RepeaterBook. Here's a direct link to the listing: **www.repeaterbook.com/repeaters/details.php?state_id=09&ID=6375.**

1. Click on **Contacts** in the **Functions** list.
2. Click on **Create Group Call.**
    - A new contact should be added to the bottom of the list.
3. Click on **Edit** next to **Contact 1** in the list.
4. Type in the name of the first TG from **Table 8.3.**

Table 8.3

W1HQ DMR Repeater Talk Group Data

	Name	Call Type	DMR ID	TS	RX Freq.	TX Freq.	Color Code
*	Worldwide	Group Call	1	1	442.100	447.100	1
	North America	Group Call	3	1	442.100	447.100	1
	Worldwide English	Group Call	13	1	442.100	447.100	1
*	TAC310 TS1	Group Call	310	1	442.100	447.100	1
	Northeast	Group Call	3172	1	442.100	447.100	1
	CT Statewide	Group Call	8901	2	442.100	447.100	1
**	CT Statewide Region 1	Group Call	8902	2	442.100	447.100	1
**	CT Statewide Region 2	Group Call	8903	2	442.100	447.100	1
**	CT Statewide Region 3	Group Call	8904	2	442.100	447.100	1
**	CT Statewide Region 4	Group Call	8905	2	442.100	447.100	1
**	CT Statewide Region 5	Group Call	8906	2	442.100	447.100	1
	TAC310 TS2	Group Call	310	2	442.100	447.100	1
	Parrot	Group Call	9998	1 or 2	442.100	447.100	1
* *PTT on demand 5 minute inactivity timeout*							
** *PTT on demand 10 minute inactivity timeout*							
Parrot allows you to transmit audio and have it played back to you ensuring call quality to adjust tour microphone gain settings.							

5. Make sure the **Group Call** is selected. If it's not, then click on the empty radio button to select **Group Call.**
6. Type the correct DMR ID for the TG into the field labeled **Code,** using information from **Table 8.3.** Click **OK** to save the information.

TG 3100

TG 3100 is the Nationwide Calling Channel and a PTT Talk Group. If the conversation continues for more than a few minutes, stations should move to TAC 310, 311 or 312 channel to clear 3100 for call in traffic only.

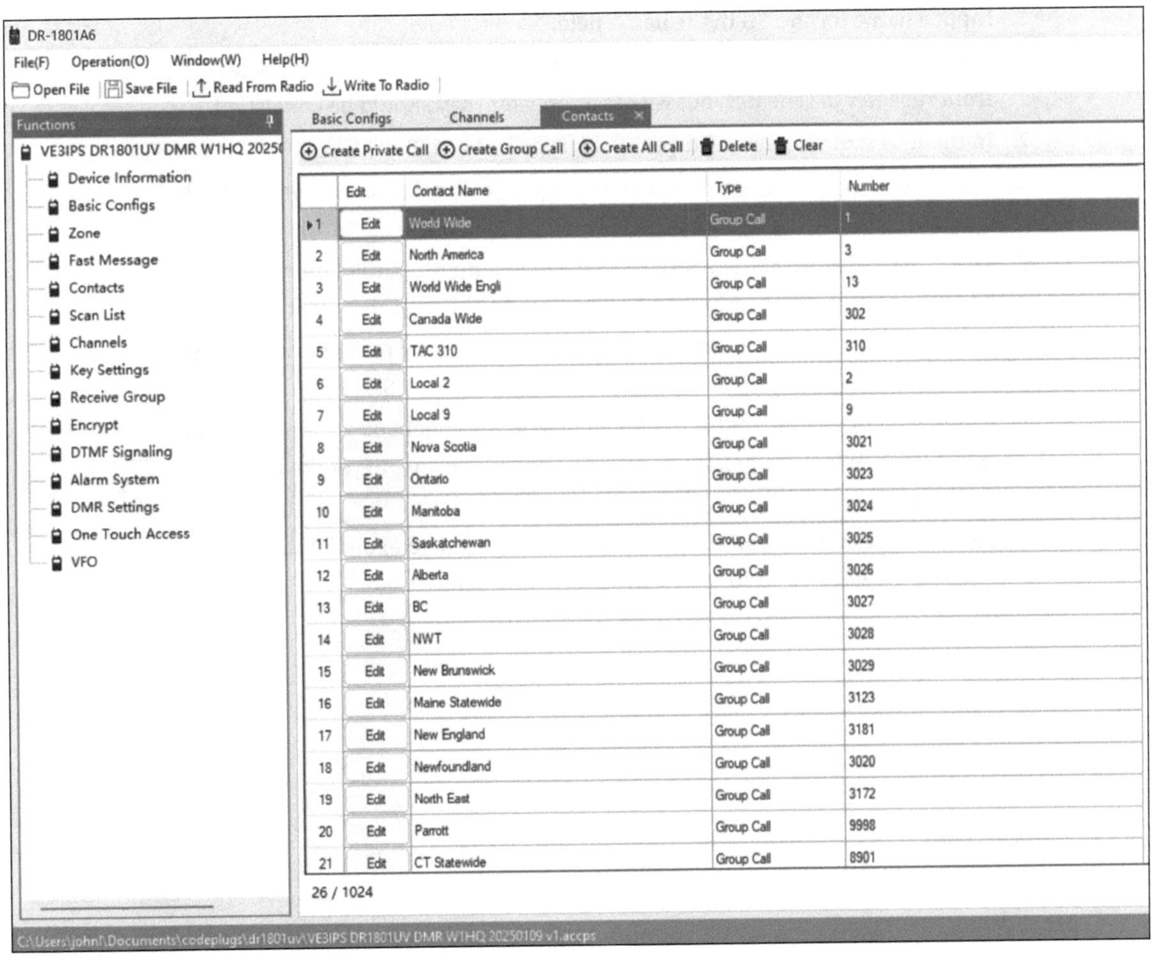

8.5.7 Define and Create Channels

Each channel includes a specific frequency, time slot, and talk group setting for a repeater. To get started, add a new channel for each frequency or TG.

1. Click on the **Channels** tab.
2. Click on **Create Channel** in the menu bar under the **Channels** tab.
3. From the dropdown menu, select **Create Digital Channel.**
   - A new channel should appear at the bottom of the channels list.
4. Click on the **Edit** button next to that new channel.
   - A popup window opens with a big **DMR** in the right-hand corner. If the window says **ANA** (for analog), you'll need to make another channel, being sure to select the digital option. Analog and digital channels have different options available.
5. Input a name for the TG in the name field.
6. In the **TX Freq. (Repeater Input)** field, type in **447.100.** These frequencies differ from repeater to repeater, but within a repeater they should be consistent.
7. In the **RX Freq. (Repeater Output)** field, type in **442.100.**
8. In the **Contact** field select the contact (TG) you would like to associate with this channel.
9. The ARRL Lab guys at the ARRL HQ have made inputting the **Color Code** simple. Slide the color code slider until a **1** shows up at the end of the slider.
10. Specify the correct **Time Slot** for the TG you are working on by moving the **Slot** slider to the left. For the W1HQ WW, that would be Time Slot 1.
11. If you want to transmit at high power, click on the checkbox beside **Tx Power** to select High. When the box is unchecked, the radio transmits at the low power setting.
12. Later on, I'll show you how to set up a **Receive Group.** to expedite that process, click on the **Receive Grouop** [sic] dropdown and select 1.

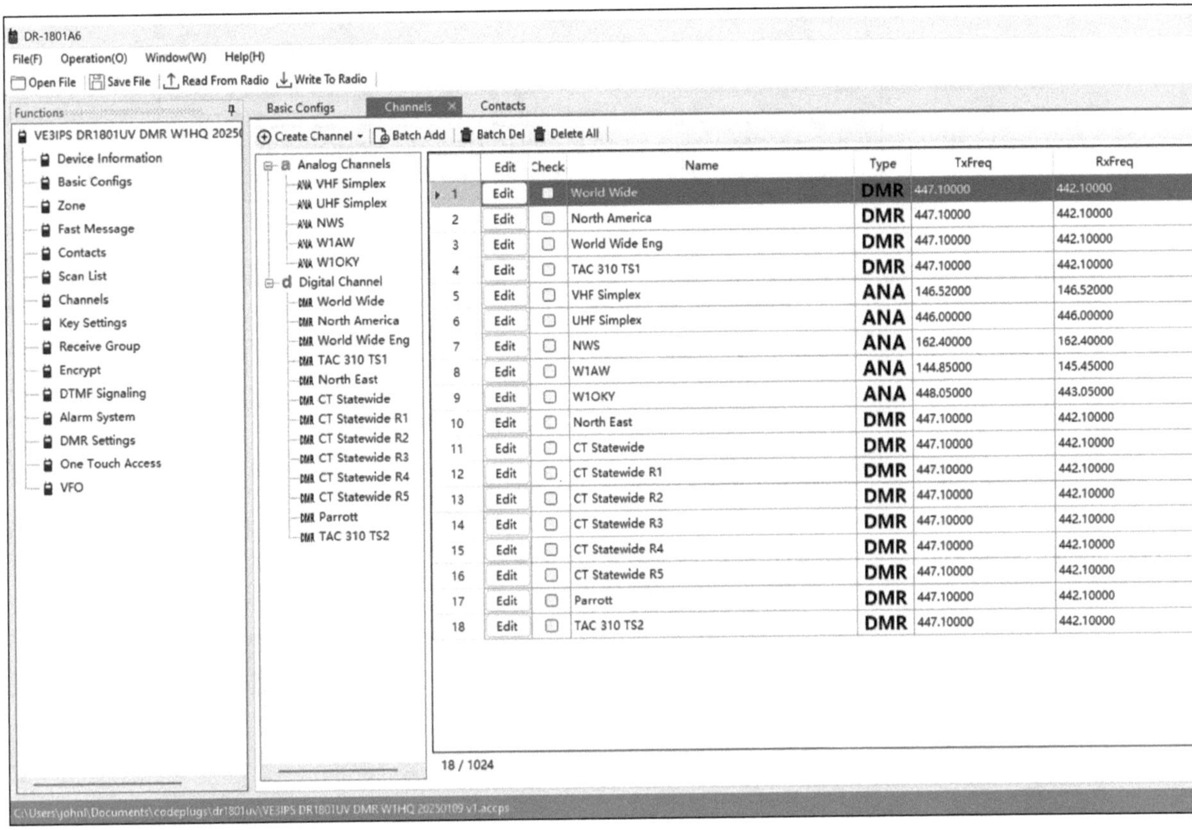

8.5.8 Organize Channels into Zones

Once you have created a channel for each of your contacts, you can start to group relevant contacts into Zones. Putting channels in zones makes navigating your radio easier. For clarity, give the zone a contextual name such a as W1HQ DMR Repeater ARES.

I assign a Zone for every repeater and name it accordingly. I also name a Zone for analog channels as well such as GTA Analog, Buffalo Analog, and ARES Simplex etc.

1. In the **Functions** list, click on **Zone**.
2. Click on the **Create** button in the menu bar in the **Zone** tab. You should see **Zone 1** appear in the Name window.
3. Type **W1HQ DMR** into the **Name** field that appears across the top of the window.
4. In the **Optional Channel** box, click on the check box beside the channel you would like to add to the **Zone Channel** box.
5. Right click on the channel with the check mark and select **Move channel to list**.
   - This will move the channel into the zone.
   - You can move multiple channels simultaneously by clicking on multiple check marks and right-clicking on one of the selected channels.

Zone Name	Channels
W1HQ	W1HQ WW TS1, W1HQ NA TS1, W1HQ WWE TS1, W1HQ TAC310 TS1, W1HQ NE TS1, W1HQ CT 1 TS1, W1HQ CT 2 TS1, W1HQ CT 3 TS1, W1HQ CT 4 TS1, W1HQ CT 5 TS1, Parrott

This is my DMR Zone set up for W1HQ:

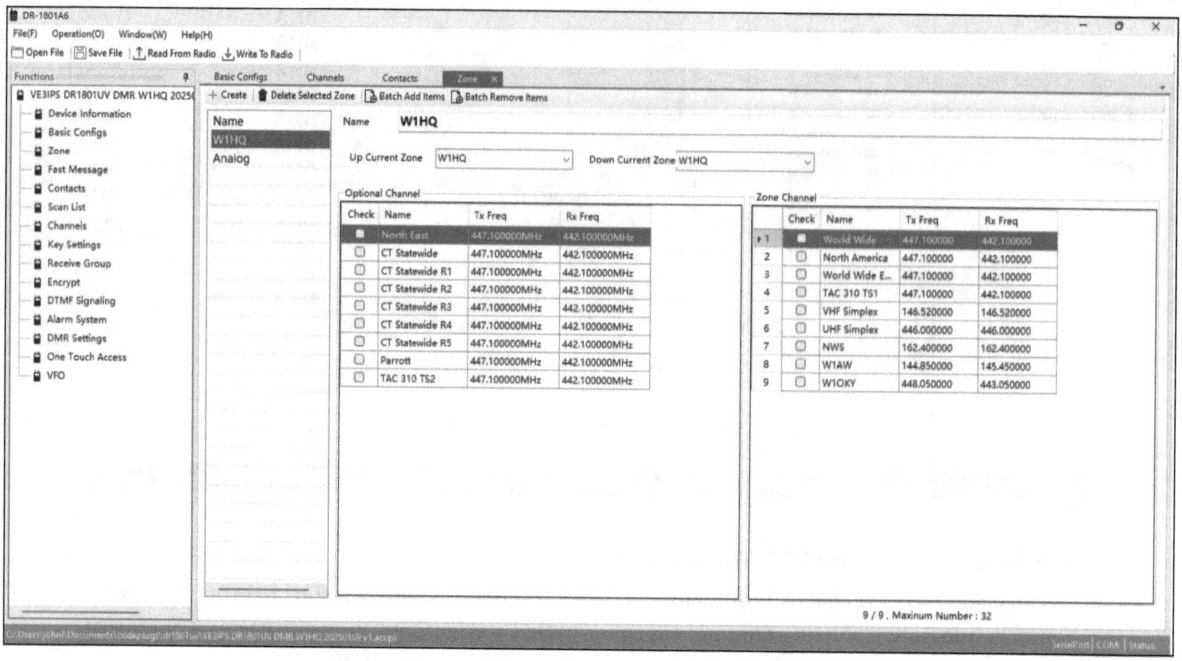

I also set up an Analog Zone using the same procedure, except that when I programmed my channels, I selected analog rather than digital.

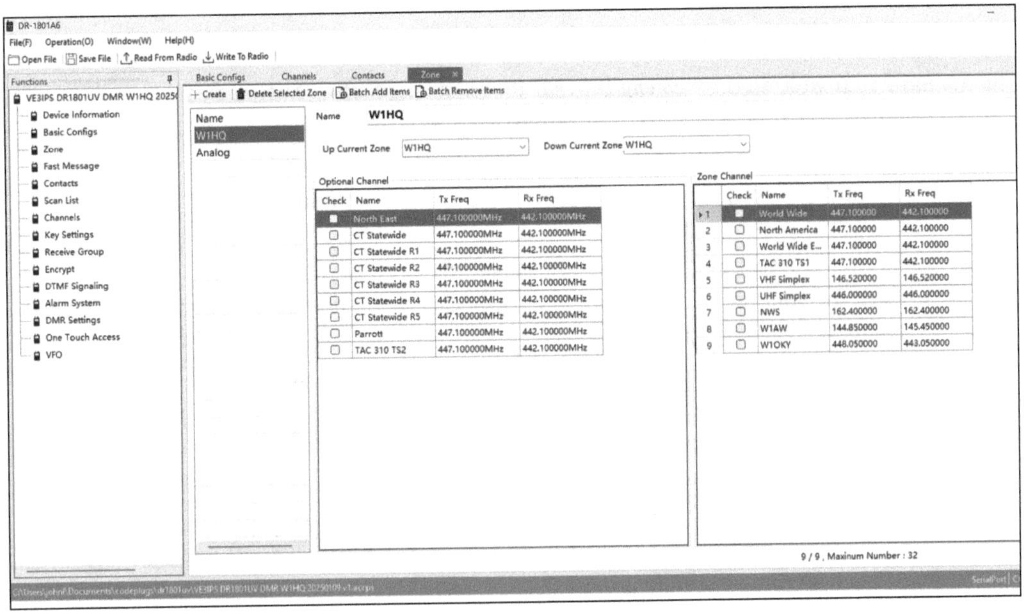

8.5.9 Program Scan Lists

If you want the radio to scan through multiple channels and stop when traffic is detected, you can set up scan lists by clicking on **Scan List** in the **Functions** pane on the left-hand side of the screen. You can create as many scan lists as you require.

Scan List Name	Channels
W1HQ Scan	All W1HQ channels

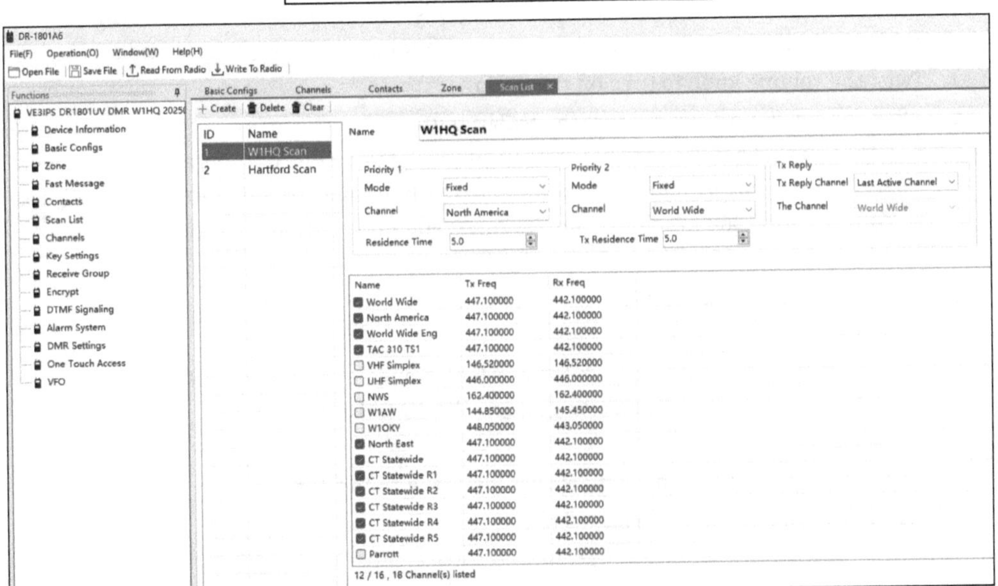

8.5.10 Save and Upload the Code Plug

1. Save your work in the *Baofeng DR-1801 Programming Software* using a contextual name like VE3IPS_DR1801UV_20250106_DMR v2.
2. Plug the programming cable into a USB port on your computer.
3. With your radio off, plug in the two-pronged speaker and mic plug into the appropriate ports on your radio.
4. Twist the volume knob until the radio turns on.
   a. Turn the volume up high enough that you can verify no one is transmitting on the frequency the radio is set to.
5. Click on the **Write To Radio** button above the **Functions** pane.
   - The LED indicator on the top of the radio will flash rapidly and a green status bar will expand onscreen while the information is uploaded to the radio.

6. The **Success** window should pop up.
7. Click **OK** to clear the window.
8. Twist the volume knob back until the radio turns off.
9. Unplug the programming cable from the radio's ports.

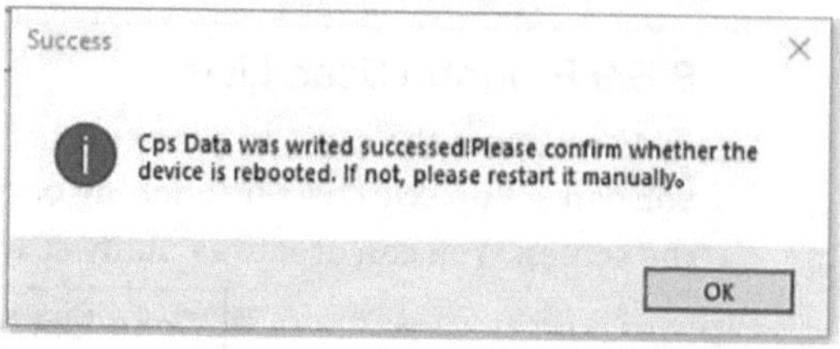

8.5.11 Test Your Code Plug

Now that you've programmed your first code plug, it's time to make sure the information transferred to the radio properly and is accurate for the repeater(s) you programmed.

1. Twist the volume knob to turn on the radio.
2. After the power up screen, press the **MENU** button.
   - Use the ▲ and ▼ buttons to navigate until the **Set** and **Zone** options appear onscreen.
   - **Set** – Allows you to adjust the radio's settings like squelch, time-out-timer, VOX, etc.
   - **Zone** – Allows you to select from the zones you programmed in order to access channels.
3. Press the ▼ key to select **Zone**.
4. Press **MENU.**
5. Use the ▼ key to navigate through the available zones, then press **MENU** to select a zone. If you've been following along, select **W1HQ DMR**.
6. Press **MENU.**
7. This should pull up the any contacts associated with the **Zone.**
8. Scroll to a contact and press **MENU** to select it.

9. This should take you back to the main screen, where you can verify the TGs. If you are within range of the repeater, you can press the **PTT** button to begin transmitting if the frequency/TG is clear.

8.6 Digital Hotspots

Digital hotspots are versatile devices that act as personal gateways (repeaters) between your Baofeng digital radio and the internet, allowing you to access worldwide networks such as DMR, D-STAR, Yaesu System Fusion (YSF), NXDN, P25, and others. These hotspots can be especially useful when you don't have access to local repeaters or want a more portable scenario for making contacts. The hotspots all use Wi-Fi to connect to your internet gateway and offer a web-based application for configuration and programming.

The hotspot acts as a gateway between your digital radio and the internet. You program your radio with the appropriate frequency and settings to communicate with the hotspot. Your radio transmits to the hotspot on a set frequency. The hotspot decodes the digital signal, processes it, and routes it to the appropriate digital network (e.g., a DMR talk group or D-STAR reflector). Incoming transmissions from the network are converted and sent to your radio.

I have several hotspots that I use and I have an openSPOT 1 that is on 24/7, a ZUMspot for portable use and an openSPOT4 PRO that I use for mobile and travel. Many times I have been in California and talking to my Toronto friends on DMR TAC312. When I was in Aruba, I used my hotspot to connect to the condominium Wi-Fi network and talk to hams around the world. I consider my hotspots to be personal repeaters allowing me to make contacts anywhere I can connect to a Wi-Fi network. The range of your Baofeng radio has just been expanded to worldwide coverage all with a hotspot that fits in the palm of your hand.

Here's an overview of popular digital hotspots like openSPOT and ZUMspot running software such as Pi-Star or WPSD:

8.6.1 SharkRF openSPOT4 Pro

The OpenSPOT4 Pro is designed for plug-and-play operation, openSPOT hotspots are excellent for users who want a device with minimal setup. This pocket friendly device has a built-in battery and software application for easy programming. The OpenSPOT4 does not require an external computer or Raspberry Pi. It includes its own hardware and web-based interface. Consider it to be like an appliance. The openSPOT4 Pro has built-in Wi-Fi and a long-lasting battery, making it a favorite for portable operation.

One of its standout features is cross-mode support. The openSPOT4 Pro's onboard transcoding hardware allows you use your Baofeng DMR transceiver to also access D-STAR, C4FM, and NXDN networks.

The openSPOT uses a Hardware AMBE Codec which provides superior audio quality compared to hotspots that rely on software AMBE emulation. Regular firmware updates add new features and improve functionality and are launched through the menu.

8.6.2 ZUMspot (with Pi-Star)

The ZUMspot is paired with a Raspberry Pi Zero board. Pi-Star is a powerful and widely used software that supports multiple digital modes, including DMR. It's based on DIY where you add software to customer supplied Linux SD cards. Its open source software allows for deep customization and community-driven enhancements. Portability can be achieved with the use of 5V USB type power banks.

8.6.3 WPSD-Based Hotspots

The new WPSD (Wireless Personal Digital System) hotspots are designed for simplicity and efficiency.

WPSD is a next-generation digital voice software suite distribution for amateur radio use. It supports the standard digital modes plus M17 (a newer digital mode) and POCSAG data/paging. WPSD is available as installable disk images, and multiple platforms & devices are supported. The WPSD Project is free and open-source software (FOSS). There is also a repeater image available.

8.6.4 DVMEGA-Based Hotspots

The core of the hotspot is the DVMEGA board with a Raspberry Pi and a Micro SD Card. With the addition of a Power Supply (110 VAC) and Antenna

The new WPSD (Wireless Personal Digital System) hotspots are designed for simplicity and efficiency.

Table 8.4

Hotspot Comparison

Feature	Pi-Star (ZUMspot)	openSPOT 4 PRO	WPSD (DVMEGA)
Ease of Use	Moderate: Requires initial setup with Pi-Star software. A learning curve for new users.	High: Plug-and-play experience with intuitive web interface and advanced auto-tuning.	Moderate: Setup through BlueDV or Pi-Star; requires some knowledge of configuration.
Customization	High: Highly customizable with Pi-Star software, including multiple modes, reflectors, and dashboards.	Moderate: Limited to the built-in SharkRF OS, but still allows reflector and network customization.	Moderate: Customizable through software but fewer options compared to Pi-Star.
Cross-Mode Support	Limited: Depends on hardware (e.g., ZUMspot doesn't natively support D-STAR-to-DMR cross-mode).	High: Excellent cross-mode support (D-STAR to DMR, C4FM to DMR, etc.).	Low: Primarily designed for single-mode or multi-mode operation without cross-mode.
Portability	High: Compact and lightweight, ideal for mobile or portable use.	High: Slim, portable, and designed for mobile operation.	Moderate: Portable, but bulkier depending on the configuration.
Price	Low: \$100–\$150 (depending on hardware version).	High: \$320–\$350 (premium device with high-end features).	Moderate: \$200–\$250 (depending on the model and accessories).

8.6.5 MMDVM-Based Hotspots

A Multi-Mode Digital Voice Modem (MMDVM) hotspot is a compact, versatile device that allows amateur radio operators to access and use various digital voice modes over the internet without relying on a local repeater. These hotspots are based on the MMDVM platform, developed by Jonathan Naylor, G4KLX, and are widely used in the ham radio community for digital voice communication. Pi-Star has been the software of choice but is being replaced with the newer WPSD software.

If you want a simple, reliable device with minimal setup, the openSPOT4 Pro is an excellent choice. If you enjoy tinkering and customizing your setup, go for a ZUMspot with Pi-Star or WPSD. The WPSD project sits in the middle as being less complex than a ZUMspot and easier to use with WPSD software. For portable use, the openSPOT4 Pro's built-in battery makes it the most convenient option.

Hotspots are an excellent solution to providing access to worldwide radio networks from the comfort of your Wi-Fi enabled home, hotel or automobile.

8.7 Getting on the Air with DMR

The Baofeng DR-1801UV is a solid upgrade for existing UV-5R users, with several standout features. The first thing you'll notice is the size of the radio, which, in my experience, is one of its strongest points. The larger form factor makes it much more comfortable to hold, and the bigger keys are easier to operate. The keyboard buttons are, however, smaller and narrower than those on the UV-82.

One notable improvement is the excellent audio quality. The upgraded speaker delivers clear and loud sound, making it a reliable choice for a variety of environments. The dual-band display is also clear and easy to read, and the buttons provide a satisfying tactile response, enhancing the overall user experience.

While the DR-1801UV introduces a refreshed design, its operation and performance remain largely similar to the UV-5R. It offers slightly more power, though the difference isn't particularly noticeable during regular use.

However, this radio is prone to some interference, especially in densely populated urban areas. In less RF-congested environments, the DR-1801UV performs reliably without major issues. In the DMR mode, the radio provides improved performance in RF-rich environments.

This radio is unlike other DMR radios that have a physical channel knob, the DR-1801UV uses the arrow and four-way controller to navigate across different channels in a talk group.

Overall, this DMR radio is a solid performer and a decent choice for the advanced amateur who is looking to expand their knowledge and experience in digital modes without breaking the bank.

8.7.1 DMR Repeater Usage Tips

Even if you're a frequent repeater user, DMR is new territory that takes some getting used to. The most important thing to know at the outset is to wait a few seconds before transmitting to allow the repeater to reset for the next transmission. To indicate that it's ready the repeater will usually play a channel free tone. Also, the DMR repeaters are linked via the Internet. A call out onto the North America talk group is received by over 350 repeaters.

Because many DMR users are either scanning or using receive groups, you should always state what talk group you are on. For example, if I were calling the Connecticut TG, I would say, "This is VE3IPS listening CT Statewide." To call into the North America TG, "This is VE3IPS listening on North America."

When calling a station, also indicate what talk group you are on, "W1AW, W1AW this is VE3IPS on CT Statewide," or "VE3RAC, VE3RAC this is VE3IPS on TAC310."

Often a user will contact another user and then move to a free talk group like TAC 310 to continue their conversation.

Do not rag-chew like you are on the HF bands or a local analog repeater on the calling talk groups. Use the TAC channels for conversations.

ID on DMR as you would on analog at least every 10 minutes.

8.8 Usability Tips for the DR-1801UV

Here are some usability tips to help you maximize your radio's potential.

8.8.1 Programming Tips

Plan your code plug:

- Define your TGs, Time Slots, and color codes for DMR operation.
- Include local analog repeaters with correct CTSS tones

Save your code plug file as a backup to make it easier to restore or modify settings later.

Input your DMR ID into the radio's settings (via programming software or menu).

Make sure you match the correct Color Code and Time Slot for your local repeater or hotspot.

8.8.2 Customize Side Keys

Assign frequently used functions (e.g., Monitor, Scan, Alarm) to the side keys for quick access. These can be configured via the menu or programming software.

8.8.3 Use Hotspots for DMR

If you don't have access to a DMR repeater, use a hotspot like a Pi-Star or openSPOT to access networks like BrandMeister or TGIF.

Program hotspot-specific talk groups to match your interests.

8.8.4 Stay Organized

Name your channels clearly for easy navigation (e.g., "Local Repeater" or "TG91 Worldwide").

Create a quick reference guide for frequently used talk groups, frequencies, or repeater settings.

8.8.5 Explore Advanced Features

Add frequently used DMR contacts for easier private calls. In this scenario, the DMR ID of the user is entered as a contact and set as Private Call.

Try out text messaging on DMR for short communications.

Organize contacts into groups for efficient communication.

8.8.6 DMR Simplex

Program the National simplex DMR channels and assign them to a DMR Simplex Zone. By convention simplex calls on DMR are made on the talk group **TG-99**, time slot **TS-1**, color code **CC-1** on the following frequencies:

- 446.075 MHz
- 433.450 MHz
- 441.000 MHz
- 446.500 MHz
- 446.075 MHz
- 433.450 MHz

8.8.7 Digital Monitor

The Baofeng DR-1801UV supports promiscuous mode, also known as open receive or digital monitor mode, which allows the radio to listen to all traffic on a specific frequency regardless of the talk group, color code, or time slot settings. This feature is useful as it will allow you to hear traffic without knowing any code plug details. Many users program SK2 with this feature.

Normally, a DMR radio filters transmissions by matching the talk group, time slot, and color code programmed into the channel. Promiscuous mode overrides these restrictions, letting you hear everything on the active frequency.

For those scanning for activity, this feature quickly allows you to assess all the active talk groups on a repeater frequency. It also allows dynamic monitoring where new or unlisted talk groups may pop up.

8.8.8 Busy Channel Lock (BCL)

BCL prevents transmission on a time slot if it is currently active. This prevents you from transmitting if the channel is active.

8.8.9 Receive Group Lists

Are an essential concept in DMR programming that determine which talk groups your radio will monitor or “listen to” on a specific channel. If you do not hear any traffic, a common problem is forgetting to put them into a list.

8.9 DR-1801UV Cheat Sheet

Menu Number	Function	Description	Options/Values
1	**Power**	Adjusts transmission power level.	High / Low
2	**Step Frequency**	Sets the tuning step size for frequency adjustments.	2.5kHz, 5kHz, 6.25kHz, 10kHz, 12.5kHz, 25kHz
3	**TX Timeout**	Sets the maximum continuous transmission time to avoid overheating.	15s, 30s, 60s, 90s, 180s, OFF
4	**Squelch**	Adjusts squelch level to block weak or unwanted signals.	Levels 0–9
5	**VOX**	Enables or disables Voice Operated Transmit and sets its sensitivity level.	OFF, Levels 1–9
6	**Scan Add/Delete**	Adds or removes channels from the scanning list.	Add / Delete
7	**DMR Color Code**	Sets the Color Code for DMR operation (equivalent to CTCSS for analog systems).	0–15
8	**DMR Repeater Slot**	Sets the timeslot for DMR communication (used for repeaters).	Slot 1 / Slot 2
9	**DMR ID**	Sets the digital DMR ID used for identification on the network.	Custom numeric entry (9 digits max)
10	**CTCS/DCS Encode**	Enables CTCSS/DCS tones for transmitting.	OFF, CTCSS (67.0–254.1 Hz), DCS (023–754)
11	**CTCS/DCS Decode**	Enables CTCSS/DCS tones for receiving.	OFF, CTCSS (67.0–254.1 Hz), DCS (023–754)
12	**Channel Mode**	Toggles between VFO (frequency) mode and Memory (channel) mode.	VFO / Channel
13	**Reverse Frequency**	Swaps transmit and receive frequencies in repeater mode for temporary testing.	ON / OFF
14	**Scan Mode**	Sets the scan mode: pausing on an active signal or continuing after a fixed time.	Time / Carrier / Search
15	**Band Width**	Sets the transmit bandwidth (wide or narrow).	Wide (25 kHz) / Narrow (12.5kHz)
16	**Busy Channel Lock**	Prevents transmission on a busy frequency.	ON / OFF
17	**Group Call ID**	Sets the Group Call ID for DMR group communication.	Custom numeric entry
18	**Private Call ID**	Sets the Private Call ID for DMR individual communication.	Custom numeric entry
19	**Keypad Lock**	Locks the keypad to prevent accidental key presses.	Manual / Auto / OFF
20	**LED Indicator**	Toggles LED status indicators for transmission, reception, and standby.	ON / OFF
21	**Voice Prompt**	Toggles voice announcements for menu navigation and operations.	ON / OFF
22	**Channel Save**	Saves the current frequency and settings to a memory channel.	Custom channel selection
23	**Reset Settings**	Resets the radio to factory default settings.	VFO Reset / Full Reset

9 Accessories for Baofeng Radios

9.1 Antennas

9.1.1 Antenna Overview

The first and most effective upgrade you should invest in for your radio is a better antenna. This one component significantly improves performance. Antenna accessories are relatively inexpensive while still offering high performance value. In my opinion, you can never have too many antennas. There's an old saying, if you can't hear the station, you can't work it.

An antenna is a device that radiates and receives electromagnetic waves, commonly used for communication systems like radios, cell phones, and Wi-Fi networks. It radiates radio signals generated by a transmitter, which propagate through the air. When connected to a receiver, the antenna captures radio waves and converts them into electrical signals for further processing. Antennas are designed to work efficiently on specific frequencies or ranges of frequencies, this frequency is known as the antenna's resonant frequency.

Basic antenna theory suggests that the ideal length for a ¼-wave antenna is 19.2 inches for VHF and 6.3 inches for UHF. Antennas are tuned to resonance for their designed frequency, generally by lengthening or shortening the radiator. When space c onstraints force an antenna to be shorter than some multiple of the resonant frequency, coils called inductors are added to bring the antenna to resonance.

Antenna gain, measured in dBi, reflects how efficiently an antenna radiates or receives energy in a specific direction compared to an isotropic radiator. The sidebar The Decibel provides a brief explanation of this rather confusing unit. For example, a 3 dBi gain effectively doubles how loud your signal sounds and improves reception. Most of these antennas have gain values ranging from 2 to 3 dBi, with longer whips offering better performance on VHF and shorter whips performing better on UHF.

The standard 5" Baofeng antenna is designed as a compromise to balance compact usability

The Decibel

The decibel (dB) is the standard unit for comparing two quantities, such as power or voltage, as a ratio. It is logarithmic so very large and very small ratios are easy to work with. Because decibels are a logarithmic value, a change of 1 dB, whether from 1 dB or 10,000 dB, represents a change of 26%. This is why 3 dB is equivalent to a doubling (a 100% increase — well, just about) of the original value. So, if your antenna has 9 dB of gain compared to an isotropic radiator, it's concentrating three times the energy toward the receiving antenna.

It's worth noting that antennas achieve gain by narrowing their radiation pattens so that a higher percentage of the transceiver's output heads toward the receiving antenna. Think about it like the fabric of an umbrella. When stretched into a dome over your head it covers a certain amount of area — hopefully enough to keep you from getting wet! — but if you pull the fabric off the frame and stand it on end it reaches out farther straight ahead (and up and down) than it does when domed. If you take that same piece of fabric and unwind it to a single thread it will extend much, much farther out in one direction, but it won't keep the rain off your shoulders at all. The same principle applies to antennas. For more on the topic, see the sidebars Effective Radiated Power and Antenna Gain in chapter 1.

Another way to increase the range of a radio signal is to put more power into the antenna. However, this is less efficient than directing the antenna's signal. Putting 1 watt into the antenna is equivalent to 30 dBm (the m means the decibels are referenced to a milliwatt) to go from 30 to 40 dBm requires 10 watts. Going from 40 to 50 dBm requires 100 watts. As you can see, at lower power levels the dBm increase is meaningful for the increase in power required, but as your power input gets higher, the return on your power input begins to diminish. On a hand-held transceiver, the jump from 1 to 5 watts, is meaningful. On a larger, permanent station, the change from a transmitter putting out 1,000 to 1,500 watts is less meaningful. Though, it should be said, there are always those who will want to push out the maximum amount of power.

and dual band functionality. Commonly referred to as rubber ducks, these antennas are constructed from flexible material and are durable enough to withstand bending and minor impacts without breaking. Overall, the rubber duck is functional but limited in range and efficiency.

For beginners in amateur radio who want a simple, out-of-the-box solution, this antenna is ideal. Its short length makes it compact and lightweight, allowing for easy storage in a pocket or backpack. While its efficiency is lower than that of upgraded antennas, it works well for accessing nearby repeaters or short-range simplex communication. High-gain whip antennas can increase your communication range, and using directional antennas like a Yagi can further enhance performance. Keeping a variety of antennas in your kit allows for better signal optimization in different scenarios.

When testing various antennas in an informal test range, I noticed improvements on transmit signals at the far end as well as better, less scratchy reception. I then tested various fringe repeaters to see if the rubber duck made for a noisy signal into the repeater, and if the gain antennas made improvements. The result was that the upgraded longer whip antennas did provide better results.

Adapters

Don't forget you may need adapters to convert the SMA-Female connector on the radio to an antenna cable. A SMA-F to BNC-F and a SMA-F to SO-239 is a mandatory accessory. Get a few of them as they are easy to lose.

The stock antenna is ideal for belt carry. I found that the 8" antennas are good replacements for the stock antenna, as their additional length and flexibility isn't a problem, even with belt carry.

I do use the longer 16" antennas for outdoor portable use, or in a fixed location due to the longer length. I have found that some of these antennas are still suitable for belt carry.

For mobile use, an antenna mounted on a magnetic base is ideal because it allows you to replace the standard antenna with one that radiates a better signal, due to the vehicle acting as a ground plane. In some cases, you can also use a longer whip antenna on a magnet mount, depending on the antenna's design.

Yagi antennas increase the maximum distance a radio signal can cover by concentrating emissions into a much narrower beam, unlike rubber duck or whip antennas. This narrowing of the antenna's radiation (or emission) pattern increases the antenna's gain. (See the sidebars Effective Radiated Power and Antenna Gain in Chapter 1.) When you want to concentrate your power in a single direction, say towards a base camp when hiking, a Yagi is a good choice. In addition, the commercially available Elk and Arrow antennas can easily be collapsed and stored in a PVC tube and then assembled quickly again when ready for use, making them popular with hikers. There are always a few repeaters I can't access with less directional antennas when I go camping. However, sticking a Yagi on a 10' to 16' mast or painter pole allows me to connect to the repeaters easily.

Table 9.1 highlights the advantages and disadvantages of different handheld antenna types, including the rubber duck antenna, longer whip antennas, a homebrew tape measure Yagi, and other manufactured handheld Yagi's.

9.1.2 Whip Antennas

Short Whips (8")

Upgrading to an 8" rubber duck antenna for the Baofeng UV-5R can significantly improve performance without being overly bulky. Be sure to choose an antenna with an SMA male connector for compatibility with the UV-5R. These antennas are still compact enough to be convenient for everyday carry.

Table 9.1

Antenna Type	Advantages	Disadvantages
Rubber Duck Antenna	- Compact and portable	- Low efficiency
	- Durable and flexible	- Limited range and gain
	- Convenient for everyday handheld use	- Poor performance in obstructed environments
Longer Whip Antenna	- Improved range over rubber duck	- Slightly larger and less portable
	- Better signal strength	- Awkward on handheld radios
	- Easy to attach and use	
Homebrew Tape Measure Yagi	- High gain and directivity	- Bulky compared to other options
	- Easy to construct and inexpensive	- Requires aiming
	- Portable and collapsible	- Less durable over time
Manufactured Handheld Yagi	- High gain and directivity	- Requires aiming
	- Lightweight models available	- Can be expensive depending on build quality
	- Excellent for satellite and weak-signal work	- Needs a boom bracket for mast mounting

Here are some of the antennas I use, all of which offer similar flexibility, performance, and cost:

- Diamond Antenna SRH519
- Nagoya NA-701
- Retevis RA-01
- Comet SMA503J

Longer Whips (16")

Upgrading to a 16" antenna for the Baofeng UV-5R can provide an even more noticeable improvement in performance compared to the standard 5" stock antenna. The increased length allows the antenna to radiate more efficiently, especially on the VHF band, where antenna length correlates closely with performance. However, the added size can make the radio bulkier and less convenient for pocket carry. This is the top choice for handheld operators needing reliable communication over greater distances.

Here are a few examples of antennas that I use:

- Signal Stuff's Super-Elastic Signal Stick: These antennas are handmade in Utah using Nitinol, a highly flexible nickel-titanium alloy, ensuring durability and longevity. Each one is individually hand-tuned for optimal resonance before being shipped.

- Nagoya NA-771, Nagoya NA-24J: These antennas are overall popular and good performers.
- Diamond SRJ77CA: This antenna is also popular and a good performer.
- Retevis's HA-01 Telescopic Antenna: This telescopic antenna is ideal for fixed use and can be collapsed to 8" for easy packing.
- ABBREE CS Tactical: This antenna's military influenced design offers a cool MIL-SPEC factor. The antenna is heavy, and its long length can be stressful to the radio antenna jack. I've found that the 18" version delivers strong performance, while the much longer 48" version tends to have poorer performance.
- N9TAX Roll-Up Slim Jim J-Pole Antenna: This antenna is ideal for increasing your range when you can hang it from a tree or pole. It's meant for portable or emergency use in a fixed location. The 60" antenna can be easily rolled up for transport and comes with a feed line.

9.1.3 Handheld Yagi and Other Directional Antennas

Tape Measure Yagi

A three-element tape measure Yagi provides significant performance improvements compared to a standard rubber duck antenna. A tape measure Yagi is a lightweight, portable directional antenna commonly used for amateur radio activities such as satellite communication, fox hunting (radio direction finding), and improving range in VHF and UHF bands. Its elements are made from retractable steel tape measures, which are flexible, durable, and easy to transport. This 2-meter band design is particularly popular among amateur radio operators because it combines functionality and simplicity. It can be easily built in just a few hours, even by beginners, and the materials—steel tape, PVC pipe, and hardware—are inexpensive and widely available. Construction typically requires only basic tools such as wire cutters, soldering equipment, and a drill.

Figure 9.1 — A dual purpose, dual boom portable tape measure Yagi for 2 meters and 70 centimeters.

The straightforward assembly process makes this antenna a great project for clubs or Scouts, with many builders successfully completing it on their first try. Detailed instructions are included in the appendix of this book.

The Yagi antenna can achieve a gain of 9 dBi, which is nearly a 6x increase in power compared to whip antennas. This gain is achieved by focusing energy in a specific direction, rather than radiating it omnidirectionally. This increases both the transmit power and receive sensitivity in the target direction. The directional nature of the antenna allows it to reach further distances, making it highly effective for simplex communication, accessing distant repeaters, or working satellites. By rejecting signals from unwanted directions—up to 20 dB from the back and sides of the antenna—the Yagi helps isolate the desired signal from interference.

Manufactured Handheld Yagi and Log Periodic Dipole Antennas

Using an Arrow Antenna, Elk Antenna, or other Yagi antenna with a Baofeng UV-5R offers significant benefits for enhancing the radio's capabilities. Antennas like the Arrow and Elk brand are specifically designed for amateur satellite work, making it easier to access satellites like SO-50 or AO-91, but are also useful for regular FM transmissions. By aiming the Yagi you can focus your signal

Understanding SWR

Look at Figure 9.A and imagine a bucket filled with water. In the center of the bucket there's a tiny motor that vibrates at a fixed rate. Let's say it oscillates back and forth at a rate of 60 times per second.

When you switch the motor on, you will see little waves moving away from the center. As the waves reach the sides of the bucket they bounce back toward the center. Of course, the motor is still vibrating, creating new waves all the time, and these will collide with the waves returning from the sides of the bucket.

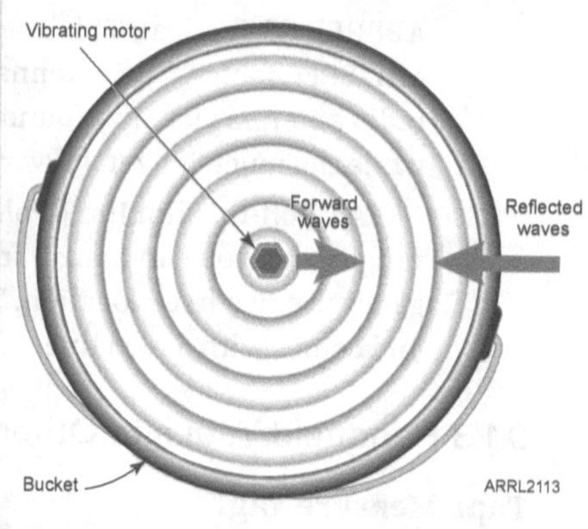

As the waves merge, they add or subtract from each other, and something strange occurs. The waves are still flowing back and forth within the bucket, but now it suddenly looks as though they are frozen in place. We call these standing waves. What appears "frozen" is the amplitude variation due to constructive and destructive interference between the two waves.

This is what happens when there is a mismatch between the impedance level your transceiver puts out and the impedance level of the antenna system. Most coaxial based antenna systems have an impedance around 50 ohms. If your transceiver is putting out 50 ohms, then the match, or SWR, is said to be 1:1 or "one to one." If your transceiver puts out 50 ohms and the antenna system has a 200 ohm impedance, then the antenna systems impedance is four times higher than the transceiver's and the SWR is said to be 4:1 or "four to one."

One to one SWR is a perfect match, while 4:1 is less than perfect. If the mismatch gets too large, then the back-flowing power can damage your transceiver. Most hams consider a mismatch of 10:1 pretty bad. A mismatch of 2 or 3:1 is generally considered ok, but in general hams strive to match their antenna systems to their transceivers as perfectly as possible.

towards distant repeaters or weak stations, improving your ability to connect over long distances. Handheld Yagi antennas are lightweight and easy to use in the field, perfect for POTA, SOTA, or urban exploration. Many are designed with easily removable elements for travel and storage. You should always maximize antenna height and minimize obstructions for best performance. A small, collapsible mast, such as a painter pole or those made by Buddipole, are the ideal choices. Please note that satellite antennas are designed for operation at 437 MHz. Using them at 446 MHz may result in an unacceptable standing wave ratio (SWR) due to the difference in design. I suggest you choose your antenna accordingly.

The following are some antennas that I regularly use:

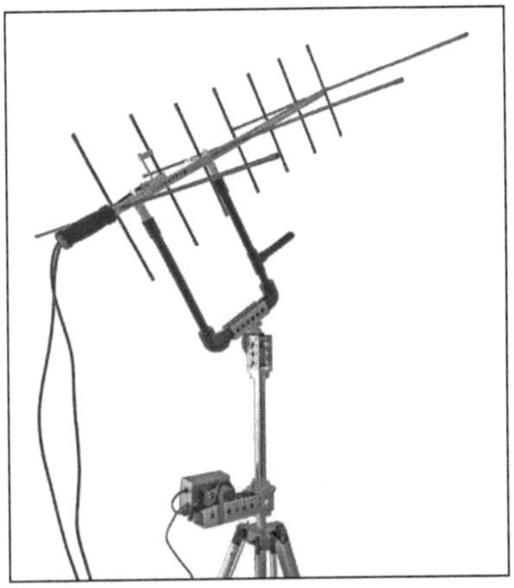

Figure 9.2 — A 2 meter and 70 centimeter Arrow Antennas Yagi mounted on a homebrew (homemade) satellite tracker.

Arrow Antenna (146/437-10)

This antenna is popular for handheld satellite communication and weak signal work. Its dual band design has separate feedlines for each band but can also use a duplexer for single radio use. The SWR is higher in the UHF FM band because the antenna is tuned for 437 MHz, but it is still acceptable for use with handheld radios. Be sure to purchase the M/BII Mounting Bracket if you plan to mount it on a mast, as it is an optional accessory. A split boom backpacker version is also available for easy packing.

Arrow also makes single band antennas with additional elements offering more gain. These are ideal for fixed use from home or for temporary deployments out in the field.

Figure 9.3 — A log-periodic Elk antenna mounted on a tripod can vastly increase your range.

Elk Antenna (2M/440L5 Log Periodic)

The Elk antenna is another popular choice, favored both as a satellite antenna and for general-purpose use. Its multi-band operation makes it ideal for portable applications. It covers the 2-meter and 70-centimeter bands in a single feedline. The log-periodic design allows for wide frequency coverage for more efficient transmissions with its low SWR across the bands.

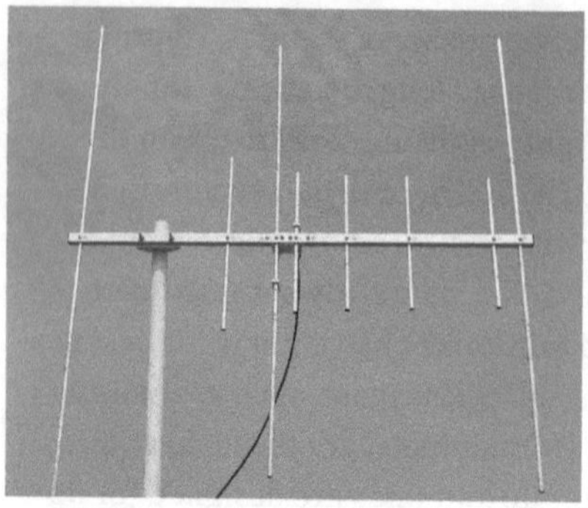

Figure 9.4 — An HYS 2 meter and 70 centimeter dual band Yagi.

HYS Tourist Band Yagi Antenna 2M/70cm

This antenna is a cost-effective model designed for non-satellite activities and is ideal for portable use where the range needs to be extended.

9.1.4 Fixed Location Antennas

These antennas are ideal for use in your home and for fixed portable operations. Due to coaxial cable losses, I suggest shorter cable runs less than 30 feet using a low-loss coaxial cable such as LMR-400. Additionally, I suggest using a jumper cable between the antenna connector and the radio, as thick coax can put a strain on the radio's antenna connector.

Arrow J-Pole

The J-pole design is a classic dual-band antenna. Its 58" length provides excellent performance, whether used in a fixed location or temporarily mounted on a 16' mast or painter's pole. It has a solid build using aluminum rods.

Comet GP-3 Dual-Band VHF/UHF Base Vertical Antenna

This antenna is mounted on a mast and is usually used in your home location. Its 6' length offers gain of 4.5 dBi (VHF) and 7.5 dBi (UHF). This high-quality base stationantenna features a one-piece, heavy-duty, UV-stabilized fiberglass radome. Despite its length, it remains easy to transport in a car. When mounted atop a 16' painter's pole, it delivers the best performance for its size.

Signal Stuff OSJ J-Pole

This antenna takes the 58" long antenna into a take-apart design that can be stored in a 20" tube, making it easy to transport or keep in a go-bag or emergency kit. I suggest adding the mast bracket to the order, as the base model is for a tripod mount. I use this mounted on a camera tripod about 5' off the ground with great results. You can place the tripod on a picnic table, vehicle, or even on top of a trailer to increase the height above ground.

9.1.5 Mobile Antennas

Magnetic Mount Mobile Antennas

A mobile antenna is an ideal choice for portable use with Baofeng radios, especially when improved performance is required. Its longer antenna length and large metal ground plane provide enhanced signal quality. These antennas are ideal for vehicle-mounted setups or temporary stationary use, offering improved range by being placed higher—such as on a car roof or an elevated surface This elevated positioning can help to avoid obstructions like buildings or terrain and provide better line-of-sight, significantly increasing transmit and receive range. Mobile antennas are also designed with higher gain than handheld antennas, making them more efficient and effective for communication over longer distances.

They can be easily moved between locations or vehicles, making them a practical option for portable operations such as field activities, road trips, or emergency scenarios. Magnetic-mount or suction-cup designs allow for quick setup and takedown without requiring permanent installation. Mobile antennas typically require a coaxial cable adapter, for example SMA-F to BNC or SO-239, to connect to the UV-5R, which adds a minor inconvenience in terms of additional components.

Magnet mounts can be placed on any metal surface that provides a suitable ground plane. I've used items like round pizza pans and even a gallon-sized tomato can.

Larsen VHF/UHF Mobile Antennas NMO2/70B

This 35" antenna offers a gain of 3.8 and 5.2 dBi and requires an NMO-type mount. Bolt-on or magnetic mounts are available. I've been using this antenna since the '80s, and its ½-wave design allows for versatile placement on your vehicle. However, depending on where you mount it, you may face challenges when trying to enter garages or underground parking lots due to its length.

Nagoya UT-72

This antenna comes complete with a strong magnetic base and is 20" long, offering a 3.5 dBi gain on UHF. It typically includes a Baofeng adapter cable in the package. I use this antenna for car rentals because of its packable size and because it fits easily into a carry-on. When packing, pay attention to where you place the magnet. I position it with the magnet at the bottom right-hand corner of my suitcase, stuck onto a small tuna can lid. Just be prepared for TSA inspections when flying.

TRAM 1185

This antenna offers a smaller 3" magnet, and its 19" length provides unity gain on VHF and 3.5 dBi gain on UHF.

Signal Stuff Super-Elastic Signal Stick

This antenna offers several magnetic mount options that utilize the handheld whip antenna for multifunction use. For the fastest, easiest "grab and go" deployment, pair it with a BNC Signal Stick and an SMA-F to BNC-F or SMA-M to BNC-F adapter. This setup allows you to have a single antenna with your radio, then easily switch it to your vehicle with the magnetic mount. It also works great at home, on top of a filing cabinet or a garage tool chest. Additionally, a 20" dual-band Signal Stalk antenna with an NMO mount is available, offering 2.5/4.5 dBi gain.

Comet CA-2x4SR BROADBAND VHF/UHF Dual Band Antenna

This antenna was designed to support Search & Rescue volunteers and professionals in the ham radio, public service, land mobile, GMRS, and FRS frequency ranges. It requires a separate mount. Comet sells several suitable options. Additionally, you'll need an SMA-F to BNC or PL-259 coaxial cable adapter. Its broadband capability offers gain performance across a wide frequency spectrum, making it a popular choice.

9.2 Power Supplies and Accessories

9.2.1 Batteries

The Baofeng UV-5R can be powered by several types of lithium-ion (Li-Ion) batteries, each offering different amperage ratings, run times, and performance characteristics. Li-Ion batteries are widely favored for their high energy density, lightweight construction, and long lifespan.

However, there are specific concerns users should be aware of to ensure safety and maximize performance. Overcharging can lead to overheating and reduce the battery's lifespan, while over-discharging may cause permanent damage. Li-Ion batteries can become unstable when exposed to extreme temperatures. Cold temperatures can temporarily reduce capacity and runtime, while heat can cause swelling or leakage. Always operate and store the battery within the manufacturer's recommended temperature range. To avoid damage, always use the manufacturer-approved charger. One important safety tip is to store spare batteries in a way that prevents their terminals from being exposed and shorted. A direct short could lead to the battery catching fire or even exploding.

Batteries have a life span that degrades over time with typical capacity being reduced after 300 – 500 charge cycles. You can extend the lifespan by avoiding frequent deep discharges. It's recommended that batteries be stored at about 50% charge capacity if not used for extended periods of time.

I check my batteries monthly, charge them as needed, and always store them in a cool, dry environment. To keep tabs, I like to mark the battery with tape and add the voltage and date that I last charged it. I have batteries that are over 5 years old and still work as specified, thanks to taking proper care of them. You should dispose of old batteries at certified recycling centers. Many electronic or hardware stores also accept them for recycling.

The current draw of the Baofeng UV-5R varies depending on its operating mode and power setting. Typically, the radio will draw about 1.5 amps at 1 watt and about 2.5 amps at 5 watts. On receiving, the current draw is around 350 mA, with around 100 mA on standby. The battery is rated at 7.4 V.

Having spare batteries is a smart move, and their affordable cost makes it easy to keep several packs on hand.

Standard 1800 mAh BL-5 Stock Battery

I've found that I get about 5 to 8 hours on a full charge, depending on usage (e.g., transmission time, standby time). For light usage, which means more listening and less talking, this typically lasts a few days.

3800 mAh Extended Battery BL-5L

This battery has nearly twice the capacity of the standard battery pack, providing around 12 to 18 hours of use, depending on transmission duty cycle. With light use, it can last about 4 days. Despite its bigger size, the radio is still pocketable, and having extended life at higher power levels is useful.

Third Party Packs

There are many third-party batteries available, and the most interesting ones are those that feature a USB-C charging cable. I'm looking to add some of these to my go-kits because they offer an easier way to charge in the field, especially with a PD power bank.

9.2.2 External Power Options

Many manufacturers now make power banks suitable for ham radio use, as the cost of batteries has decreased and there's more interest in having the ability to charge or to run power devices during outages. On long trips or during field operations, you can use a portable power bank to charge the UV-5R through its USB charging port (available on certain models). Some pure sine wave power banks with an AC 120 V outlet receptacle also allow you to use the 120 V charging cradle.

You can also connect the UV-5R to 12 V dc power supplies for extended operations. Some dc adapters come with a vehicle utility port cord and a dummy battery pack insert to connect directly to the radio. These items are inexpensive but add a tremendous capability when needed. Power supplies rated for at least 5 amperes are ideal for this purpose. The Powerwerx PA-12-6A compact power supply is a good choice.

9.3 Cases and Other Equipment

9.3.1 Cases

There are several types of cases available for Baofeng radios, including protective cases, carry bags, and holsters. These cases provide protection for the radios, convenient portability, and even additional storage for accessories. They're available online from third party suppliers in any colour, design, or carry style.

The soft silicone case is molded to fit the radio snugly, offering good shock protection. While it doesn't provide waterproofing or extreme protection, it's suitable for everyday use. I've found that it does need occasional cleaning. These cases come in different colors, which can help identify the owner, indicate intended usage, or simply reflect personal style. For example, one of my fellow ARES operators uses an orange case for all his emergency radios.

Plastic or nylon cases are also useful for protecting the radio's display and buttons. Some of these cases offer enhanced belt clips for added strength and ruggedness.

A solution I've found helpful is using a zip case or small multi-purpose tactical carry bag to store everything. About the size of or smaller than a lunch bag, it can hold all the radio accessories, allowing me to treat it like a go-kit or for EDC use. This grab-and-go option is ideal for busy operators.

Hard plastic cases like Pelican, Meiju, and Nanuk brands are also ideal for carrying your radio equipment. The foam-filled inserts need to be plucked out, but these cases' waterproof capability is ideal for protecting your equipment in rugged conditions.

Carry cases are personal. With a little experimentation, you can decide what works best for you.

9.3.2 Controllers, Programmers and TNCs

Below are some interesting add-ons to enhance your experience and increase your usability of your Baofeng radio:

TidRadio BL-1 Bluetooth Programming Adapter

Baofeng does not manufacture an official Bluetooth programming adapter specifically for the UV-5R. However, third-party manufacturers like TidRadio produce Bluetooth programming adapters that are compatible with the UV-5R and other Baofeng radios. These adapters plug into the radio's 2-pin speaker/microphone jack and connect to your smartphone, tablet, or computer via Bluetooth, eliminating the need for a USB programming cable. Bluetooth-compatible apps like CHIRP and OD Master can be used for programming.

G1LRO Universal Radio Controller

The G1LRO Universal Radio Controller (URC) is an accessory designed to make programming and operating radios like the Baofeng more convenient, especially for controlling the radio through a computer. The ability to use the URC as a computer interface for digital modes presents the ability to send packet and even SSTV images across your radio.

The URC was designed to allow a user to connect their Baofeng to the controller and be able to do three things:

1. USB-C interface for power and sound card interface
2. Push-to-Talk (PTT) using the CM108 standard or through a serial port
3. Serial Programming interface to connect to CHIRP programming software

The interfaces include a TRRS jack for receiver audio and PTT, as well as OHIS (Open Headset Interface Standard) jacks for connecting to a computer or Heil-brand communication headsets. This is an ideal setup for using your radio in a base station.

Argent Data Systems

For APRS, a TNC is required. The OpenTracker USB bundle includes a cable for connecting to the radio and power (Baofeng Style, CT-H5), along with an ADS-GM1 GPS receiver. Just add a UV-5R radio, and you're ready to go. For initial setup, you'll need a standard USB cable to program the radio on your Windows PC and input your call sign and settings.

The TNC takes the GPS location data and transmits it in APRS format. This data can be received directly by individual APRS-enabled radios or through APRS I-gates (which connect an APRS radio to the internet for passing data for mapping), making it viewable on the aprs.fi website.

I suggest dedicating a UV-5R specifically for this task. It can be installed in a vehicle for real-time tracking, but others also use it for tactical purposes. For example, in a community event, APRS can help track assets, like a command post vehicle, ensuring everyone knows their location.

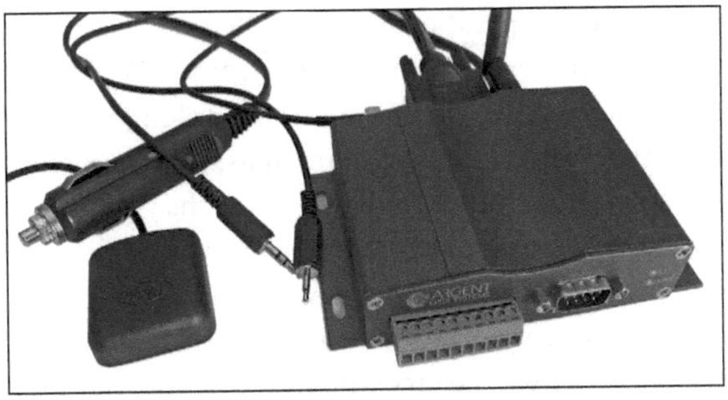

Figure 9.5 — The Argent Data Systems Tracker4 APRS Tracker/TNC/Gateway. This is an older model than the one written about in the text.

Byonics

The Byonics TinyTrak4 is an APRS tracker controller, digipeater, and KISS TNC. It offers enhanced capabilities that other TNCs do not. When connected to a serial GPS and a radio, it can send and receive APRS position and telemetry packets, as well as act as a standalone smart digipeater. A computer running APRS software can be connected to the second serial port to provide mapping of received traffic. You will need to purchase their GPS5 GPS receiver and the HTK2 power/interface cable.

Figure 9.6 — The Byonics TinyTrak4.

The TinyTrak3Plus bundle is a simpler APRS GPS position encoder. When connected to the GPS5 GPS and a UV-5R, it will transmit its location at an adjustable rate, providing an inexpensive way to build a mobile tracker.

I use a dedicated TinyTrak inside an ammo box, along with an antenna and power source, for APRS tracking out in the field.

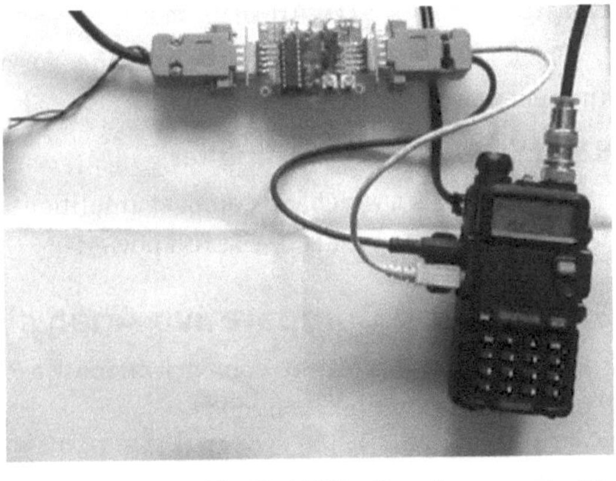

Figure 9.7 — The TinyTrak3 Plus Bundle connected to the Baofeng UV-5R.

9.3.3 Power Amplifiers

Using a power amplifier with the Baofeng UV-5R can significantly enhance its transmission power, enabling greater distances and improving communication performance. This is particularly beneficial in areas with high interference or long-range operations. The UV-5R typically outputs around 4 to 5 watts by default, but with a power amplifier this can increase from 20 W to 50 W, or even more depending on the amplifier used. This increase in power allows for longer-range communication and better signal penetration, especially in dense urban areas, environments with heavy interference, or challenging terrain such as hills or buildings. It turns your handheld radio into a powerful mobile or base station configuration. For mobile operations, like ARES or field activations such as POTA (Parks on the Air), the amplifier ensures that your signal reaches farther and remains clear.

Amplifiers work best when paired with high-quality antennas. If your antenna isn't efficient, the amplifier will be less effective, and you may end up with poor signal quality despite the extra power. I suggest mobile or base station antennas as described, as rubber duck antennas are not suitable for amplifier usage.

Mirage BD-35

The Mirage BD-35 dual-band amplifier boosts your handheld radio to a much more powerful 45 watts on 2-meters or 35 watts on 440 MHz. You can get 35 watts out with just 4 watts in. It features power protection, an "On Air" indicator light, and a rugged metal case, giving your signal a big punch when needed.

BTECH AMP-V25 VHF Amplifier

The BTECH amplifier stands out for its easy integration with Baofeng radios. It simply connects to the Baofeng SMA antenna jack using the included jumper cable. You can also connect an external microphone, making it useful for both mobile and base station setups. It allows for up to 40 watts output and even features a built-in speaker. This is quite an interesting product, but it's important to note that it's only for the VHF band.

RM Italy LA-144 VHF Amplifier

The LA-144 is a 60 W 12 V linear amplifier for use in the VHF 135 – 175 MHz band, requiring an input power of 0.5 – 4 W maximum.

RM Italy LA-435 UHF Amplifier

The LA-435 is a wideband compact amplifier for the UHF band, covering from 405 to 480 MHz. Output is nominal 60 W at full power.

9.4 Baofeng Comparative Analysis

Table 9.2 shows a feature comparison for the Baofeng UV-5R, UV-82, and DR-1801UV, highlighting key aspects for each model.

The table shows that the UV-5R is best for beginners on a budget, as it offers essential features in a compact, affordable package. The UV-82, by comparison, is suitable for operators who need better audio, higher power output, and a larger form factor and dual PTT, offering a slightly more rugged design. And the DR-1801UV suits those venturing into DMR, as it offers digital and analog modes along with DMR-specific features like text messaging and contact storage.

Table 9.2

Baofeng	UV-5R	UV-82	DR-1801UV
Frequency Range	VHF: 136-174 MHz, UHF: 400-520 MHz	VHF: 136-174 MHz, UHF: 400-520 MHz	VHF: 136-174 MHz, UHF: 400-470 MHz
Power Output	1W / 4W	1W / 5W	1W / 5W
Channels	128 Channels	128 Channels	3,000 Channels
Digital Mode (DMR)	No	No	Yes (Tier I and Tier II)
Analog Mode	Yes	Yes	Yes
Dual-Band Operation	Yes	Yes	Yes
Dual Watch / Dual PTT	Dual Watch, Single PTT	Dual Watch, Dual PTT	Dual Watch, Dual PTT
Battery Capacity	1,800 mAh	2,800 mAh	2,200 mAh
Programming Software	*CHIRP*-compatible	*CHIRP*-compatible	*CPS* (Customer Programming Software)
Audio Quality	Basic	Improved	High clarity (digital mode)
Speaker Output Power	1W	1W	1W
Display	Standard LCD	Enhanced LCD	Standard LCD
Repeater Capability	Yes (Analog)	Yes (Analog)	Yes (Digital & Analog)
Text Messaging	No	No	Yes (DMR only)
Color Code and Timeslot	No	No	Yes (for DMR)
Contact Storage	Limited	Limited	1000+ contacts (for DMR)
Antenna Type	Removable SMA-F	Removable SMA-F	Removable SMA-F
Emergency Alert Function	Yes	Yes	Yes
Battery Life (Typical Use)	Up to 12 hours	Up to 15 hours	Up to 14 hours (longer in DMR mode)
Weight (w/ Battery)	204 g	214 g	222 g
Dimensions (HxWxD)	110 x 58 x 32 mm	126 x 60 x 36 mm	130 x 60 x 40 mm
Price Range (USD)	$	$$	$$$

Table 9.3
Handheld Transceivers to Expand Your Capabilities
Digital Transceivers

Manufacturer	Model	Description	
Kenwood	**TH-D75A**	144/220/440 MHz band TX/RX 5 W max output APRS and D-STAR compatible DV fast data mode Built in GPS module Simultaneous reception on two digital channels Wideband multimode reception	
Yaesu	**FT-70DR**	108 – 580 MHz RX, 144/430 MHz TX 5W max output C4FM FDMA (System Fusion) compatible Digital Group Monitor feature	
Icom	**ID-52A**	144/430 MHz TX/RX 5 W max output D-STAR compatible Bluetooth® features enabled Simultaneous VHF/UHF RX Photo TX/RX	
Connect Systems Inc.	**CS7000 M17**	400 – 512 MHz TX/RX M17 compatible* DMR compatible* FCC Part 90 compliant IP67 waterproof rating Updatable firmware	
Anytone	**AT-D878U-VII**	144/430 MHz TX/RX 5 W max output Bluetooth and GPS 10,000 talk group memory 500,000 digital contacts memory	

***Requires a firmware update**

(continued)

Analog Transceivers

Manufacturer	Model	Description	
Yaesu	**VX-6R**	. 5 – 999 MHz RX 144/220/430 MHz TX 5 W max output Submersible 89 Short wave broadcast channels preloaded 64 special DTMF memories 16-digit, 10-memory DTMF autodialer	
Icom	**IC-V86 (or IC-U86)**	144 MHz TX/RX 7 W max output 1500 mW audio IP54 rated MILI-STD 810G rated Simultaneous VHF/UHF RX Photo TX/RX	 ▲ IC-V86  ▲ IC-U86
Alinco	**DJ-VX50T**	144/430 MHz TX/RX 5 W max output Aircraft and weather band RX IP67 rated Color display	

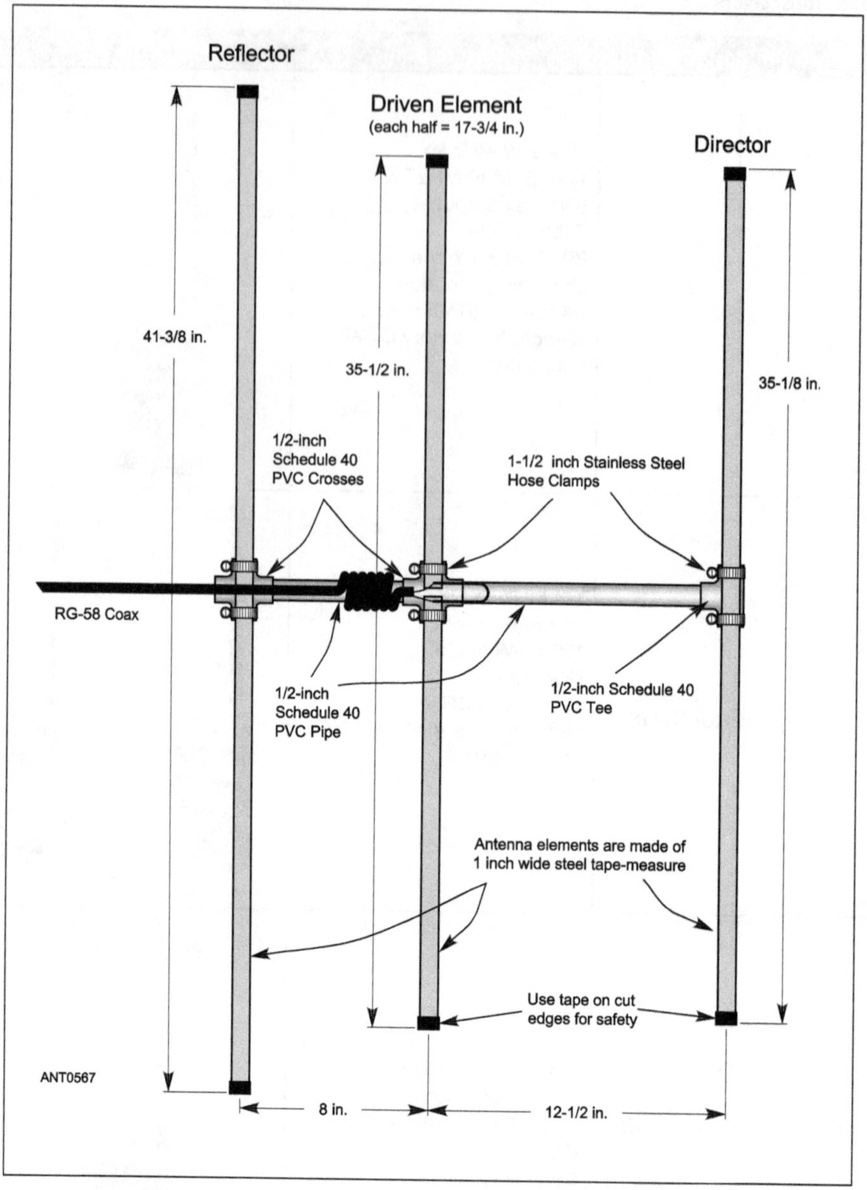

Tape measure Yagi schematic.

Glossary

Words in definitions that are *italicized* have a separate glossary entry. A word in **bold** is defined in that entry.

73 — Ham abbreviation for "best regards." Generally expressed at the end of a contact.

Adapters — Connectors that convert one type to another.

Allocations — Frequencies authorized for a particular FCC telecommunications service.

Alternating current or voltage (ac) — Electrical *current* or *voltage* with a direction or polarity, respectively, that reverses at regular intervals.

Amateur operator — A person named in an amateur operator/primary license station grant on the ULS consolidated licensee database to be the control operator of an amateur station.

Amateur Radio Emergency Service (ARES®) — An organization of amateur volunteers that is sponsored by the ARRL and provides emergency communication services to groups such as the American Red Cross and local Emergency Operations Centers (EOC).

Amateur service — A radio communication service for the purpose of self-training, intercommunication and technical investigations carried out by amateurs, that is, duly authorized persons interested in radio technique solely with a personal aim and without *pecuniary* interest.

Amateur station — A station licensed in the amateur service, including necessary equipment, used for amateur communication.

American Radio Relay League (ARRL) — The national association for Amateur Radio.

Ampere (A) — The basic unit of electrical *current,* also abbreviated amps. One ampere is the flow of one *coulomb* of charge per second.

Amplifier — A device or piece of equipment used to *amplify* a signal.

Amplify — Increasing the strength or *amplitude* of a signal.

Amplitude — The strength or magnitude of a signal.

Amplitude modulation (AM) — The process of adding information to a signal or *carrier* by varying its amplitude. Transmissions referred to as AM phone are usually composed of two sidebands and a carrier. Shortwave broadcast stations use this type of AM, as do stations in the Standard Broadcast Band (535-1710 kHz). AM in which only one sideband is transmitted is called single-sideband or SSB and is the most popular voice mode on the *high frequency (HF)* bands.

AMSAT (Radio Amateur Satellite Corporation) — Organization that manages many of the amateur satellite programs.

Analog (linear) signal — A signal (usually electrical) that can have any amplitude (voltage or current) value, and whose amplitude can vary smoothly over time. Also see *digital signal.*

Antenna — A device that radiates or receives radio frequency energy.

Apogee — The point in a satellite's orbit at which it is farthest from the Earth. See perigee.

Automatic Packet Reporting System (APRS) — A system by which amateurs report their position automatically by radio to central servers from which their locations can be observed.

Amateur Radio Direction Finding (ARDF) — Competitions in which amateurs combine orienteering with *radio direction finding.*

Antenna analyzer — A portable instrument that combines a low-power signal source, a frequency counter and an *SWR meter.* Also known as an SWR analyzer.

Array — An antenna with more than one *element*. In a driven array all *elements* are *driven elements.* In a parasitic array some *elements* are *parasitic elements.*

Attenuate — To reduce the strength of a signal. An attenuator is a device that attenuates a signal.

Audio frequency (AF) signal — An ac electrical signal in the range of 20 hertz to 20 kilohertz (20,000 hertz). This is called an audio signal because human hearing responds to sound waves in the same frequency range.

Automatic control — A station operating under the control of devices or procedures that ensure compliance with FCC rules.

Autopatch — A device that allows users to make telephone calls through a *repeater.* Auxiliary station — A station that operates in support of another station, such as a *repeater,* by transmitting control information or relaying audio.

Balanced line — *Transmission line* in which none of the conductors is connected directly to *ground.*

Balun — Contraction of "balanced to unbalanced" and pronounced "BAHL-un." A device to transfer power between a balanced load and an unbalanced feed line or device, or vice versa.

Band — A range of frequencies. An **amateur band** is a range of frequencies on which amateurs are allowed to transmit.

Band plan — Voluntary organization of activity on an amateur band under normal circumstances.

Bandwidth — (1) Bandwidth is the range of frequencies occupied by a radio signal. (2) FCC Part 97 defines bandwidth for regulatory purposes as "The width of a frequency band outside of which the mean power is attenuated at least 26 dB below the mean power of the transmitted signal within the band." [Part 97.3 (8)]

Base — A station at a fixed location.

Battery — A package of one or more *cells.*

Battery pack — A package of several individual *cells* connected together (usually in series to provide higher voltages) and treated as a single battery.

Baud — The rate at which symbols are transmitted in a *digital mode.*

Beam antenna — See directional antenna.

Block diagram — A drawing using boxes to represent sections of a complicated device or process. The block diagram shows the connections between sections. A block diagram shows the internal functions of a complex piece of equipment without the detail of a *schematic diagram.*

BNC — A type of RF connector.

Bonding — Connecting equipment or circuits together to keep them at the same voltage.

Break-in — Switching between transmit and receive during CW operation so that you can listen to the operating frequency between Morse *elements* (full break-in) or during short pauses in your transmissions (semi-break-in).

Breaking in — The term for joining an ongoing contact by transmitting your *call sign* during a pause in the contact.

Broadcasting — One-way transmissions intended to be received by the general public, either direct or relayed.

Bug — A mechanical Morse *key* that uses a spring to send dots automatically.

Call — (1) Abbreviated form of *call sign.* (2) Attempt to make contact.

Call district — The ten administrative areas established by the FCC.

Call sign — The letters and numbers that identify a specific amateur and the country in which the license was granted.

Calling frequency — A frequency on which amateurs establish contact before moving to a different frequency. Usually used by hams with a common interest or activity.

Capacitance — A measure of the ability to store energy in an *electric field.* Capacitance is measured in *farads.*

Capacitor — An electrical *component* that stores energy in an *electric field.* Capacitors are made from a pair of conductive surfaces called electrodes that are separated by an *insulator* called the dielectric.

Carrier — The unmodulated RF signal to which information is added during *modulation.* Also see *modulate.*

Cell (electrochemical cell) — A combination of chemicals and *electrodes* that converts chemical energy into electrical energy. See *battery.*

Centi (c) — The metric prefix for 10^{-2} or division by 100.

Certificate of Successful Completion of Examination (CSCE) — A document that verifies that an individual has passes one or more exam *elements.* A CSCE is good for 365 days and may be used as evidence of having passed an *element* at any other amateur license exam session.

Channel — A range of *frequencies* used for one radio or communications signal.

Channel spacing — The difference in frequency between *channels.*

Charge — Store energy in a *battery* by reversing the chemical reaction in its *cells.*

Chassis ground — The common connection for all parts of a *circuit* that connect to the metal enclosure or chassis (pronounced "CHAA-see") of the circuit.

Check in — Register your station's presence on a *net* with the *net control station.*

Choke balun — A type of *balun* made by forming a coaxial feed line into a coil or placing ferrite cores on the feed line, creating an *RF choke.*

Circuit — A conductive path through which *current* can flow.Circuit breaker — A protective component that "breaks" or opens a *circuit* or *trips* when an excessive *current* flow occurs.

Citizen's Band (CB) —An unlicensed radio service operating near 27 MHz intended for use by individuals and businesses over ranges of a few miles.

Closed repeater — A *repeater* that restricts access to members of a certain group of amateurs. See *open repeater.*

Closed circuit — An electrical circuit with an uninterrupted path for the current to follow. Turning a switch on, for example, closes or completes the circuit, allowing current to flow. Also called a complete circuit.

Coaxial cable — Coax (pronounced KOH-aks). A type of *transmission line* with a single center conductor inside an outer shield made from braid or solid metal and both sharing a concentric central axis. The outer conductor is covered by a plastic jacket.

Common — Term for the shared reference for all voltages in a circuit. Also referred to as *circuit* common. See *ground* and *bonding.*

Common-mode — Currents that flow equally on all conductors of a multiconductor cable, such as speaker wires or telephone cables, or on the outer surface of shielded cables.

Common-mode choke — see *choke balun.*

Communications emergency — A situation in which communication is required for immediate safety of human life or protection of property.

Component — (1) A device having a specific quantity of an electrical property (such as *resistance*) or that has a specific electrical function. (2) One signal of a group that makes up a composite signal.

Conductor — A material in which *electrons* move freely in response to an applied *voltage.*

Connector — A *component* used to connect and disconnect electrical circuits and equipment.

Continuous wave (CW) — Radio communications transmitted by on/off *keying* of a continuous radio-frequency signal. Another name for international *Morse code.*

Control code — Information in the form of data or tones used to adjust a station under *remote control.*

Control link — The means by which a control operator can make adjustments to a station operating under *remote control.*

Control operator — The person designated by the licensee of a station to be responsible for the transmissions of an amateur station.

Control point — The location at which the control functions of the station are performed.

Controlled environment — Any area in which an RF signal may cause radiation exposure to people who are aware of the radiated electric and magnetic fields and who can exercise some control over their exposure to these fields. The FCC generally considers amateur operators and their families to be in a controlled RF exposure environment to determine the *maximum permissible exposure* levels. See *uncontrolled environment.*

Conventional current — See *current.*

Core — In an *inductor,* the core is the material or space the wire is wound around or passed through.

CORES — Commission Registration System of the *FCC.*

Coulomb (C) — The basic unit of electrical charge. One coulomb is 6.25 × 1018 *electrons.* 1 *ampere* equals the flow of 1 coulomb of electrons per second.

Courtesy tone (beep) — A short burst of audio transmitted by a *repeater* to indicate that the previous station has stopped transmitting. It can also be used to indicate that the *time-out timer* has been reset.

CQ — "Calling any station," the general method of requesting a contact with any station.

Crossband — Able to receive and transmit on different amateur frequency bands. For example, a *repeater* might receive a signal on 70 cm and retransmit it at 2 meters.

CTCSS — Continuous Tone Coded Squelch System. A low frequency tone (also called subaudible tone) required to access many *repeaters.* See *PL.*

Current (electrical) — The movement of electrons in response to an electromotive force, also called *electronic current. Conventional current* is the flow of positive charge that moves in the opposite direction of electronic current.

Cycle — One complete repetition of a repeating waveform, such as a *sine wave.*

CW (Morse code) — See *continuous wave.*

Data (digital) mode — Computer-to-computer communication, such as by *packet radio* or FT8, in which information is exchanged as data characters or digital information.

DC voltage — A voltage with a constant *polarity.* See *direct current.*

Deceptive (or false) signals — Transmissions that are intended to mislead or confuse those who may receive the transmissions. For example, distress calls transmitted when there is no actual emergency are false or deceptive signals.

Decibel (dB) — In electronics decibels are used to express ratios of *power, voltage,* or *current.* 1 dB = 10 log10 (power ratio) or 20 log10 (voltage or current ratio).

Deci (or lower case d) — The metric prefix for 10–1 or division by 10.

Degree — A measure of angle or phase. There are 360 degrees in a circle or *cycle.*

Demodulate — To recover the information from a modulated signal by reversing the process of *modulation.* See *modulate.*

Detect — (1) To determine the presence of a signal. (2) To recover the information directly from a modulated signal.

Detector — The stage in a receiver in which the *modulation* (voice or other information) is recovered from a *modulated* RF signal.

Deviation — The change in *frequency* of an *FM carrier* due to a *modulating* signal. Also called carrier deviation.

Diffraction — To alter the direction of a radio wave as it passes by edges of or through openings in obstructions such as buildings or hills. Knife-edge diffraction results if the dimensions of the edge are small in terms of the wave's wavelength.

Digipeater — A type of *repeater* station that retransmits or forwards digital messages.

Digital mode — See *data mode.*

Digital signal — (1) A signal (usually electrical) that can only have certain specific amplitude values, or steps — usually two; 0 and 1 or ON and OFF. (2) See *data mode.*

Digital signal processing (DSP) — The process of converting an *analog signal* to *digital* form and using a microprocessor to process the signal in some way such as filtering or reducing noise.

Diode — An electronic *component* that allows electric *current* to flow in only one direction.

Diplexer — A device that allows radios on two different bands to share a single antenna. Diplexers are used to allow a dual-band radio to use a single dual-band antenna. See *duplexer.*

Dipole— As used in Amateur Radio, the term usually refers to a *half-wave dipole* antenna.Direct conversion — A type of receiver that recovers the modulating signal directly from the modulated RF signal.

Direct current (dc) — Electrical *current* that flows in only one direction.

Directional antenna — An antenna with an ability to receive and transmit that is enhanced in a specific (forward) direction and attenuated in one or more directions. See *front-to-back ratio* and *front-to-side ratio.*

Director — A *parasitic element* of a *Yagi antenna* that focuses the radiated signal in the desired direction. See *reflector.*

Discharge — Extract energy from a *battery* or *cell*. Self-discharge refers to the internal loss of energy without an external *circuit.*

Dish — A curved *directional antenna* that uses a *reflector* to focus radio waves.

Distress call — A transmission made in order to attract attention in an emergency.See *MAYDAY* and *SOS.*

Doppler shift — A change in observed frequency of a signal caused by relative motion between the transmitter and receiver. Also called the Doppler effect.

Doubling — Two or more operators transmitting at the same time on the same frequency.

Downlink — Transmitted signals or the range of frequencies for transmissions from a satellite to Earth. See *uplink.*

Driven element — An antenna *element* supplied directly with power from the transmitter.

Dual-band antenna — An antenna designed for use on two different amateur bands.

Dummy antenna or dummy load — A station accessory that dissipates a transmitted signal as heat to allow testing or adjustment of transmitting equipment without radiating a signal on the air.

Duplex — (1) Transmitting on one frequency and receiving on another, such as for repeater operation. (2) A mode of communications (also known as **full duplex**) in which a user transmits on one frequency and receives on another frequency simultaneously. This is in contrast to **half duplex** in which the user transmits at one time and receives at other times.

Duplexer — A device that allows bidirectional communication on closely spaced frequencies or *channels*. In a *repeater,* the duplexer also allows the transmitter and receiver to share a single antenna. See *diplexer.*

Duty cycle — The percentage of time that a signal or device, such as a transmitter, is active. **Duty factor** is the same as duty cycle, but expressed as a fraction instead of percent.

DX — Distance, distant stations, foreign countries.

DXpedition — An expedition for the purpose of making contacts from a rare or unusual location.Earth connection — An electrical connection to the Earth for electrical safety purposes. See also *ground.*

Earth station — An amateur station located on or within 50 km of the Earth's surface, intended for communications with space stations or with other Earth stations by means of one or more other objects in space.

Earth-Moon-Earth (EME) or moonbounce — A method of communicating with other stations by reflecting radio signals off the Moon's surface.

Echolink — A system of linking repeaters and computer-based users by using the Voice-Over-Internet Protocol.

Electric field — A region of space in which electrical energy is stored and in which a stationary electrically charged object will feel a force. The **electric potential** between two points in the electric field is the amount of energy required to move a single electron between those two points.

Electrode — The general term for an electrical contact or connection point.

Electromagnetic wave — Energy composed of a continuously varying *electric field* and *magnetic field* moving through space or a *transmission line.*

Electromotive force (EMF) — The force that causes *electrons* or other charged objects to move.

Electron — A negatively charged atomic particle. Moving electrons make up an electrical *current.*

Electronic current — See *current.*

Element — (1) The conducting part or parts of an antenna designed to radiate or receive *radio waves.* (2) An examination for an FCC license in the amateur service. (3) A dot or dash in the *Morse code.*

Elmer — A ham radio mentor or teacher.

Emcomm — An abbreviation for *emergency communications.*

Emergency — A situation where there is an immediate threat to the safety of human life or property.

Emergency communications — Communications conducted under adverse conditions where normal channels of communications are not available.

Emergency traffic — Messages with life and death urgency or requests for medical help and supplies that leave an area shortly after an emergency.

Emission — The transmitted signal from an *amateur station.*

Emission privilege — Permission to use a particular *emission type* (such as Morse code or voice).

Emission types — Term for the different modes authorized for use on the amateur radio bands. Examples are CW, SSB, FT8, and *FM.*

Encoding — Changing the form of a signal into one suitable for storage or transmission. **Decoding** is the process of returning the signal to its original form.

Encryption — Changing the form of a signal into a privately-known format intended to obscure the meaning of the signal. **Decryption** is the process of reversing the encryption.

Energy — The ability to do work; the ability to exert a force to move some object.

Envelope — The outline of an RF signal formed by the peaks of the individual RF cycles.

Farad (F) — The basic unit of *capacitance.*

Federal Communications Commission (FCC) — Federal agency in the United States that regulates use and allocation of the frequency spectrum among many different services, including Amateur Radio.

Federal Registration Number (FRN) — An identification number assigned to an individual by the FCC to use when performing license modification or renewal.

Feed line — See *transmission line.*

Feed line loss — The fraction of power dissipated as heat as it travels through a feed line.

Feed point — The point at which a *transmission line* is electrically connected to an antenna.

Feed point impedance — The ratio of RF voltage to current at the *feed point* of an antenna.

Ferrite — A ceramic material with magnetic properties used in *inductors*. Ferrite is often formed into beads or cores so that it may be placed on cables, forming an *RF choke*.

Forward power — Power in a *transmission line* traveling from a transmitter toward a *load* or *antenna*.

Fox hunting — Exercises in which participants look for a hidden transmitter (the fox) to test *radio direction-finding* skills. Also called a **bunny hunt**.

Frequency — The number of complete cycles per second of an *ac current* or *ac voltage*.

Frequency band — A continuous range of frequencies in which one type of communications is authorized. See *band*.

Frequency coordination — Allocating *repeater* input and output frequencies to minimize interference between repeaters and to other users of the band.

Frequency coordinator — An elected individual or group that recommends repeater frequencies to reduce or eliminate interference between *repeaters* operating on or near the same frequency in the same geographical area.

Frequency discriminator — A *detector* used for *FM* signals.

Frequency modulation (FM) — The process of adding information to an RF signal or *carrier* by varying its frequency. FM broadcast stations and most professional communications (police, fire, taxi) use FM. **FM phone** is used on most *repeaters*.

Frequency privilege — Authorization to use a particular group of frequencies.

Frequency-shift keying (FSK) — A method of digital *modulation* that shifts the transmitter frequency to represent the bits of digital data.

Front-end overload — Interference to a receiver caused by a strong signal that causes the receiver's sensitive input circuitry ("front end") to be overloaded. Front-end overload results in distortion of the desired signal and the generation of unwanted spurious signals within the receiver. See *receiver overload*.

Front-to-back ratio (F/B) — The ratio of an antenna's *gain* in the forward direction to that in the opposite or rear direction.

Front-to-side ratio (F/S) — The ratio of an antenna's *gain* in the forward direction to that at right angles to the forward direction.

FRS — Family Radio Service. An unlicensed radio service that uses low-power radios operating near 460 MHz and intended for short-range communications by family members.

Fundamental — The frequency of which all *harmonics* are integer multiples.

Fundamental overload — *Radio frequency inter-ference (RFI)* caused when a strong RF signal exceeds a receiver's ability to reject it.

Fuse — A thin metal strip mounted in a holder. When excessive *current* passes through the fuse, the metal strip melts and opens the *circuit* to protect it against further current overload.Gain — (1) Enhancing an antenna's ability to receive or radiate signals in a specific direction. (2) The ability of a component, circuit, or piece of equipment to *amplify* a signal. (3) **Mic Gain** — sensitivity of the microphone amplifier circuit. (4) **RF Gain** — sensitivity of the receiver to incoming RF signals. (5) **AF Gain** — receiver audio output volume.

Gateway — A station that serves to connect one network of stations with the internet or another network of stations.

Giga (G) — The metric prefix for 10^9 or multiplication by 1,000,000,000.

GMRS — General Mobile Radio Service. A licensed radio service operating 460 MHz intended for family businesses and members to communicate within a city or region.

Go-kit — A pre-packaged collection of equipment or supplies kept at hand to allow an operator to quickly report where needed in time of need.

Ground — (1) An electrical connection to the Earth (*Earth connection*) for the purposes of electrical safety. (2) A reference voltage point or surface (see *ground plane*) in a circuit or system of equipment. (3) To connect a circuit to a ground.

Ground loss — RF energy that is converted to heat while reflecting from or traveling through or along the Earth's surface.

Ground rod — A metallic rod driven into the Earth to make a ground connection or *Earth connection*. Also known as a ground electrode or earth electrode.

Ground plane — A conducting surface of continuous metal or discrete wires that acts to create an electrical image of an antenna. **Ground-plane antennas** require a ground-plane in order to operate properly.

Ground-wave propagation — *Propagation* in which radio waves travel along the Earth's surface.

Ham-band receiver — A receiver designed to receive only frequencies in the amateur bands.

Hamfest — A flea-market for ham radio, electronic and computer equipment and accessories.

Half-wave dipole — A popular antenna that is 1⁄2-wavelength long at the desired operating frequency. Dipoles usually consist of a single length of wire or tubing with a feed point at the center. See *dipole*.

Harmful interference — Interference that seriously degrades, obstructs or repeatedly interrupts a radio communication service operating in accordance with the Radio Regulations. [Part 97.3 (a) (22)]

Harmonic — A signal that is an integer multiple (2×, 3×, 4×, etc) of a *fundamental* frequency.

Handheld radio — A VHF or UHF transceiver that can be carried in the hand or pocket. Also known as an **HT**.

Health and Welfare traffic — Messages about the well-being of individuals in a disaster area. Such messages must wait for **Emergency** and **Priority traffic** to clear, and results in advisories to those outside the disaster area awaiting news from family and friends.

Henry (H) — The basic unit of *inductance*.

Hertz (Hz) —The basic unit of frequency. 1 Hz = 1 cycle per second.High frequency (HF) — The term used for the frequency range from 3 MHz to 30 MHz.

ICS (Incident Command System) — method of organizing a response to emergencies and disasters used in the United States. See also *National Incident Management System*.

Impedance — The ratio of ac voltage to ac current, including *phase*. The combination of *reactance* and *resistance* that constitutes opposition to ac current. Impedance is measured in *ohms*.

Impedance match — To adjust *impedances* to be equal or the case in which two impedances are equal. Usually refers to the point at which a feed line is connected to an antenna or to transmitting equipment. If the impedances are different, that is a mismatch. See *matched*.

Inductance — A measure of the ability to store energy in a *magnetic field*. Inductance is measured in *henries*.

Inductor — An electrical *component* that stores energy in a *magnetic field*. An inductor is usually composed of a coil of wire wound around a central core.

Input frequency — A *repeater's* receiving frequency.

Insulator — A material in which *electrons* do not move easily in response to an applied *voltage*. Insulation is used to prevent *current* flow between points at different voltages.

Intermodulation (intermod or IMD) — Spurious signals created by the combination of other signals. Usually related to the overload of circuits by strong signals.

International Amateur Radio Union (IARU) — The international organization of national Amateur Radio societies.

International Telecommunication Union (ITU) — The organization of the United Nations responsible for coordinating international telecommunications agreements.

Internet Radio Linking Project (IRLP) — A system of linking repeaters by using the Voice-Over-Internet Protocol.

Inverted V — A *dipole* antenna supported in the middle with each half sloping downward.

Ion — An electrically-charged atom or molecule.

Ionosphere — A region of electrically charged (ionized) gases high in the atmosphere that affects the *propagation* of radio waves through it.

Isotropic antenna — An *antenna* that radiates and receives equally in all directions, both vertical and horizontal.

Keplerian elements — Numeric parameters describing a satellite's orbit that can be used to compute the position of the satellite at any point in time.

Kerchunk — The sound made when a brief transmission activates a repeater.

Keyboard-to-keyboard — A digital mode intended for operators to exchange text messages as the characters are entered.

Kilo (k) — The metric prefix for 10^3 or multiplication by 1000.

Light-emitting diode (LED) — A diode that emits light when current flows through it.

Line-of-sight propagation — The term used to describe VHF and UHF *propagation* within the radio horizon in a straight line directly from one station to another.

Linear — (1) To act on a signal such that the result is a replica of the original signal at a different scale. (2) Equipment that amplifies the output of a transmitter, often to the full legal amateur power limit of 1500 W *peak envelope power (PEP)*.

Liquid-crystal display (LCD) — A device for displaying graphics or characters by passing light through a liquid crystal between patterns of *electrodes*.

Local control — Operation of a station with a *control operator* physically present at the transmitter.

Load — A device or system to which electrical *power* is delivered, such as a heating element or *antenna*. Also the amount of power consumed or that can be safely dissipated, such as a "50-watt load."

Loading — (1) Attaching or increasing an electrical load. (2) Increasing an antenna's apparent electrical length by inserting inductance or capacitance.

Lobe — A direction of maximum reception or transmission in an antenna's *radiation pattern*. The main lobe has the greatest strength for the entire pattern. A **side lobe** is a maximum located at an angle to the main lobe.

Log — The documents of a station that detail operation of the station. They can be used as supporting evidence and for troubleshooting interference-related problems or complaints.

Log periodic — a type of frequency independent antenna.

Loop — (1) An antenna with *element(s)* constructed as continuous lengths of wire or tubing. (2) A point of maximum voltage or current on an antenna.

Low Frequency (LF) — 30 to 300 kHz. A reference to the 2200 meter band.

Lower sideband (LSB) — (1) In an AM or *single sideband* signal, the *sideband* located below the *carrier* frequency. (2) The common single sideband operating mode on the 40, 80, and 160 meter amateur bands.

Machine — Slang for *repeater*.

Magnetic field — A region of space in which magnetic energy is stored and in which a moving electrically charged object will feel a force.

Matched — A transmission or feed line that is terminated by a load that has the same *impedance* as the feed line's *characteristic impedance*.

Maximum permissible exposure (MPE) — The maximum intensity of RF radiation to which a human being may safely be exposed. FCC Rules establish maximum permissible exposure values for humans to RF radiation. [Part 1.1310 andPart 97.13 (c)]

MAYDAY — From the French *m'aidez* (help me), MAYDAY is used when calling for emergency assistance in voice modes.

Medium Frequency (MF) — 300 kHz to 3 MHz. A reference to the 630 meter or160 meter bands.

Mega (M) — The metric prefix for 10^6 or multiplication by 1,000,000.

Memory channel — Frequency and mode information stored by a radio and referenced by a number or alphanumeric designator.

Meteor scatter — Communication by signals reflected by the ionized meteor trails in the upper atmosphere.

Metric prefixes — A series of terms used in the *metric system* of measurement.

Metric system — A system of measurement that uses a set of prefixes to indicate multiples of 10 of a basic unit.

Medium frequency (MF) — The term used for the frequency range from 300 kHz to 3 MHz.

Micro (μ) — The metric prefix for 10^{-6} or division by 1,000,000.

Microphone (mic or mike) — A device that converts sound waves into electrical energy.

Microwave — The conventional term for frequencies greater than 1000 MHz (1 GHz).

Milli (m) — The metric prefix for 10^{-3} or division by 1000.

Mobile station — Any station that can be operated while in motion, typically in a car, but also on a boat, a motorcycle or bicycle, truck or RV.

Mobile flutter — Rapid amplitude variation of a signal from a moving vehicle experiencing multipath interference. Also called picket fencing.

Mode — (1) The combination of a type of information and a method of transmission. For example, *FM* radiotelephony or *FM phone* consists of using *FM* modulation to carry voice information. (2) The combination of a satellite's *uplink* and *downlink* bands.

Mode-restricted — Portions of the amateur bands in which only certain *emission types* are allowed.

Modem — Short for modulator/demodulator. A modem changes data into audio signals that can be transmitted by radio and *demodulates* a received signal to recover transmitted data.

Modulate — The process of adding information to an RF signal or *carrier* by varying its *amplitude, frequency,* or *phase.*

Morse code — The system of encoding characters as dots and dashes invented by Samuel Morse. See *continuous wave.*

Multiband antenna — An antenna capable of operating on more than one amateur band, usually using a single *transmission line.*

Multimode radio — Transceiver capable of SSB, CW and *FM* operation.

N or type N connector — A type of RF *connector* that can be used through *microwave* frequencies.

Nano (n) — The metric prefix for 10^{-9} or division by 1,000,000,000.National Electrical Code (NEC) — A set of guidelines governing electrical safety, including antennas.

National Incident Management System (NIMS) — The method by which emergency situations are managed by US public safety agencies.

Net — A formal system of operation in order to exchange or manage information.

Net control station (NCS) — The station in charge of a *net.*

Network — (1) A term used to describe several digital stations linked together to relay data over long distances. (2) A general term for electrical circuits.

Node — (1) One station in a digital network. (2) A point of minimum voltage or current on an *antenna.*

Noise blanker — A circuit or function that mutes the receiver output during noise pulses.

Noise reduction — Removing random noise from a receiver's audio output.

Null — (1) Tune or adjust for a minimum response (2) A direction of minimum reception or transmission in an antenna's *radiation pattern.*

Offset frequency — The difference between a *repeater*'s transmitter and receiver frequencies. Also known as the repeater's **split** or **offset**.

Ohm — The basic unit of electrical *resistance* and represented by the symbol Ω.

Ohm's Law — A basic electrical law stating that the *current* (I) through a *circuit* is directly proportional to the *voltage* (E) across the circuit and inversely proportional to the *resistance* (R) of the circuit: I = E / R. Ohm's Law is equivalently stated asE = I × R or R = E / I.

Omnidirectional — An *antenna* that radiates and receives equally in all horizontal directions.

One-way communications — Radio signals not directed to a specific station, or for which no reply is expected. The FCC Rules provide for limited types of one-way communications on the amateur bands. [Part 97.111 (b)]

Open circuit — A break in an electrical *circuit* that prevents *current* from flowing.

Open repeater — A *repeater* available for use by all hams.

Operator/primary station license — An amateur license actually consists of two licenses. The operator license is that portion of an Amateur Radio license that gives permission to operate an amateur station. The primary station license is that portion of an Amateur Radio license that authorizes an *amateur station* at a specific location. The station license also lists the *call sign* of that station.

OSCAR — Orbiting Satellite Carrying Amateur Radio

Oscillate — To vibrate continuously at a single *frequency*. An **oscillator** is a device or circuit that generates a signal at a single frequency.

Output frequency — A *repeater*'s transmitting frequency.

Packet radio — A system of digital communication using the AX.25 protocol whereby information is broken into data groups called **packets** that also contain addressing and error-detection information.

Paddle —A pair of contacts operated by one or two levers used to control an electronic keyer that generates *Morse code* automatically.

Parallel circuit — An electrical *circuit* in which *current* may follow more than one path.

Parasitic element — An antenna *element* that affects the antenna performance by receiving and re-radiating energy from a *driven element* without being connected directly to the feed line.

Part 15 — The section of the FCC's rules that deal with unlicensed devices likely to transmit or receive RF signals.

Part 97 — The section of the FCC's rules that regulate Amateur Radio.

Peak envelope power (PEP) — The average *power* during one RF cycle of a radio signal at the crest of the modulated *waveform*.

Pecuniary — Payment of any type, whether money or other goods or services.

Perigee — The point in a satellite's orbit at which it is closest to the Earth. See *apogee*.

Period — The time it takes for one complete cycle of a repeating waveform. The reciprocal of *frequency*.

Phase — A measure of position in time within a repeating waveform, such as a sine wave. Phase is measured in *degrees* or radians.

Phase modulation (PM) — The process of adding information to a signal by varying its *phase*. Phase modulation is very similar to *FM*. PM signals can be received by *FM* receivers.

Phase-shift keying (PSK) — A method of digital *modulation* that shifts the transmitted signal's phase to represent the bits of digital data.

Phone — Another name for voice communications. An abbreviation for radiotelephone.

Phone emission — The FCC's name for voice transmissions.

Phone patch — Conducting a telephone call via radio communications.

Phonetic alphabet — A standardized list of words used on voice modes to make it easier to understand letters of the alphabet, such as those in call signs. The call sign KA6LMN stated phonetically is Kilo Alfa Six Lima Mike November.

Pico (p) — The metric prefix for 10^{-12} or division by 1,000,000,000,000.

PL (see CTCSS) — An abbreviation for Private Line, a trademark of Motorola.

Plug — An electrical *connector* designed to be inserted into a *jack* or *receptacle.*

Polarity — The orientation or direction of a *voltage* or *current* with respect to a convention that assigns positive and negative.

Polarization — The orientation of the *electric field* of a *radio wave.* A radio wave can be horizontally, vertically, or circularly polarized.

Portable designator — Additional identifying information added to a call sign specifying the station's location.

Portable device — Generally considered to be a radio transmitting device designed to be transported easily and set up for operation independently of normal infrastructure. For purposes of RF exposure regulations, a portable device is one designed to have a transmitting antenna that is generally within 20 centimeters of a human body.

Potential — see *voltage.*

Potentiometer — (pronounced po-ten-chee-AH-me-ter) Another name for a variable *resistor* in which the resistance value can be changed without removing it from a circuit. Also called a **pot**.

Power — The rate of energy consumption or expenditure. To calculate power in an electrical *circuit* multiply the *voltage* applied to the circuit by the *current* through the circuit ($P = I \times E$).

Power amplifier — See *linear.*

Power density — The strength of a *radio wave* measured as power per unit of area.

Power supply — A device that converts ac power from a utility or other service to ac or dc power used by equipment.

Prefix — The leading letters and numbers of a call sign that indicate the country in which the *call sign* was assigned.

Primary service — When a frequency band is shared among two or more different radio services, the primary service is preferred. Stations in a **secondary service** must not cause harmful interference to, and must accept interference from stations in the primary service. [Part 97.303]

Primary station license — See *operator/primary station license.*

Priority traffic — Emergency-related messages, but not as important as *emergency traffic.*

Privileges — The frequencies and modes of communication that are permitted in an FCC telecommunications service.

Procedural signals (prosign) — For *Morse code* communications, one or two letters sent as a single character, such as AR or SK, to indicate the operator's intention or to control the communication. For *phone* communications, prosigns consist of single words, such as "Break" or "Over."

Propagation — The method by which *radio waves* travel.

Protocol — A method of *encoding,* packaging, and exchanging digital data.

Push to talk (PTT) — Turning a transmitter on and off manually with a switch, usually thumb- or foot-activated.

Q signals — Three-letter symbols beginning with Q used in *Morse code* to save time and to improve communication. Some examples are QRS (send slower) and QTH (location).

Q system — A method of providing signal quality reports on a scale of 1 ("Q1") to5 ("Q5").

QSL card — A postcard that serves as a confirmation of communication between two hams. QSL is a *Q-signal* meaning "received and understood."

QSO — A conversation between two radio amateurs. QSO is a *Q signal* meaning"I am in contact."

Quad antenna — A *directional antenna* with *elements* in the shape of four-sided loops, one *wavelength* in circumference.

Quarter-wave vertical — A *ground-plane* antenna constructed of a 1⁄4-wavelength radiating *element*, usually oriented perpendicularly to the Earth or ground-plane.

Question pool — The set of questions from which an amateur license exam is constructed. There is one pool for each license class.

Radial — A wire forming part of a *ground plane*, attached at an antenna's base and running radially away from the antenna.

Radiation — To emit or give off energy, such as a radio wave. **Ionizing radiation** has sufficient energy to cause an electron to escape from an atom, creating a charged *ion*. RF energy used for radio communication is much less energetic and is called **non-ionizing radiation**.

Radiation pattern — A graph showing how an *antenna* radiates and receives in different directions. An **azimuthal pattern** shows radiation in horizontal directions. An **elevation pattern** shows radiation at different vertical angles.

Radio Amateur Civil Emergency Service (RACES) — A part of the Amateur Service that provides radio communications for civil defense organizations during local, regional or national civil emergencies.

Radio direction finding (RDF) — The method of locating a transmitter by determining the bearings of received signals.

Radio frequency (RF) exposure — FCC Rules establish maximum permissible exposure (MPE) values for humans to RF radiation. [Part 1.1310 and Part 97.13 (c)]

Radio frequency (RF) signals — RF or **radio signals** are generally considered to be any electrical signals with a frequency higher than 20,000 Hz, up to 300 GHz.

Radio-frequency interference (RFI) — Disturbance to electronic equipment or to radio communication caused by radio-frequency signals.

Radiogram — A formal message exchanged via radio.

Radio horizon — The most distant point on the Earth to which radio signals can travel without *ionospheric* or tropospheric propagation.

Radio wave — An *electromagnetic wave* with a frequency greater than 20 kHz.

Ragchew — An informal conversation.

Range — The distance over which radio signals are exchanged.

Reactance — The property of opposition to ac current. Capacitors exhibit **capacitive reactance** and inductors exhibit **inductive reactance**. Reactance is measured in *ohms*.

Receiver (RVCR) — A device that converts radio waves into signals we can hear, see, or be read by a computer.

Receiver overload — Interference to a receiver caused by a RF signal too strong for the receiver input circuits. A signal that overloads the receiver RF input causes front-end overload. Receiver overload is sometimes called **RF overload**.

Recharge — See *charge*.

Rectify — Convert *ac* to *dc*.

Reflected power — Power in a *transmission line* returning to the transmitter from the load or antenna.

Reflector — (1) A *parasitic element* of a *Yagi antenna* that cancels the radiated signal in the undesired direction. (2) A conducting surface that acts as an electrical mirror to reflect radio waves.

Refract — Bending of an *electromagnetic wave* as it travels through materials with different properties. Radio waves are refracted as they travel through the ionosphere.

Region — Administrative areas defined by the *International Telecommunication Union* (ITU).

Remote control — Operation of a station in which the control functions of the station are operated by a *control operator* over a *control link*.

Remote receiver — A receiver at a separate location from a transmitter. Used by *repeater* systems to extend listening range or by individual stations to improve reception capabilities.

Repeater — A station that retransmits the signals of other stations to give them greater range.

Resistance — The property of opposing an electric *current.* Resistance is measured in ohms.

Resistor — An electronic *component* with a specific value of *resistance,* used to oppose or control current through a *circuit.* Resistors can be either fixed or variable. (See *potentiometer.*)

Resonance — The condition in an electrical circuit or antenna in which *reactance* is zero.

Resonant circuit — A circuit that exhibits *resonance* at one or more frequencies.

Resonant frequency — The frequency at which a circuit or antenna is resonant.See *tuned circuit.*

RF burn — A burn produced by contact with high RF voltages.

RF choke — An *inductor* or other impedance used to prevent or reduce the flow ofRF current.

RF connector — A type of electrical *connector* designed specifically for use withRF signals.

RF feedback — Distortion of transmitted speech caused by RF signals being picked up by the microphone input circuits.

RF ground — The technique of maintaining the enclosures of radio equipment at a common RF voltage. See *bonding.*

RF overload — See *receiver overload.*

RF safety — Preventing injury or illness to humans from the effects of radio-frequency energy.

Rubber duck antenna — A flexible rubber-coated antenna used mainly with hand-held VHF or UHF transceivers.

S meter — A meter that provides an indication of the relative strength of received signals in S-units.

Safety ground — A *ground* connection intended to prevent shock hazards.

Scanning — Rapidly switching between frequencies to listen for an active *channel.*

Scattering — Radio wave *propagation* by means of multiple reflections in the layers of the atmosphere or from an obstruction.

Schematic diagram — A drawing that describes the electrical connections in a piece of electric or electronic equipment by using symbols to represent the electrical *components.*

Schematic symbol — A standardized symbol used to represent an electrical or electronic *component* on a *schematic diagram.*

SDR (Software Defined Radio) — Radio in which signal processing functions are performed by software that can be reconfigured automatically or under operator control.

Secondary service — See *primary service.*

Selectivity — The ability of a receiver to distinguish between signals.

Semiconductor — (1) A material with conductivity between that of a *conductor* and an *insulator.* (2) An electrical *component* constructed from semiconductor material.

Sensitivity — The ability of a receiver to detect signals.

Series circuit — An electrical *circuit* in which there is only one path for the *current* to follow.

Service — A set of regulations by the FCC that defines a certain type of telecommunications activity.

Shack — The room or location in which an *amateur station* is constructed.

Shield — (1) A cable's metallic layer or coating intended to prevent external signals from being picked up by an internal conductor or to prevent signals from being radiated from the internal conductor. (2) A metal wall or case that blocks RF signals.

Shielding — Surrounding an electronic circuit with conductive material to block RF signals from being radiated or received.

Short circuit — An electrical connection that causes *current* to bypass the intended path. Short-circuit often refers to an accidental connection that results in improper operation of equipment or circuits.

Sideband — An RF signal that results from modulating the amplitude or frequency of a *carrier*. An AM sideband can be either higher in frequency (**upper sideband** or **USB**) or lower in frequency (**lower sideband** or **LSB**) than the carrier. *FM* sidebands are produced on both sides of the carrier frequency.

Signal generator — A device that produces a low-level signal that can be set to a desired frequency.

Signal report — An evaluation of the transmitting station's signal and reception quality. See *Q system*.

Simplex — Receiving and transmitting on the same frequency. See *duplex* and *half-duplex*.

Sine wave — A *waveform* with an amplitude equal to the sine of frequency × time.

Single sideband (SSB) — SSB is a form of *amplitude modulation* in which one *sideband* and the *carrier* are removed.

Skyhook — Slang for antenna.

SMA — A type of RF connector used at *microwave* frequencies.

Software Defined Radio — see *SDR*.

SOS — A *Morse code* call for emergency assistance.

Space station — An *amateur station* located more than 50 km above the Earth's surface.

Speaker — A device that turns an *audio frequency* electrical signal into sound.

Specific absorption rate (SAR) — A term that describes the rate at which RF energy is absorbed by the human body. *Maximum permissible exposure (MPE)* limits are based on whole-body SAR values.

Spectrum — The range of electromagnetic signals. The radio spectrum includes signals between audio frequencies and infrared light.

Splatter — A type of interference to stations on nearby frequencies that occurs when a transmitter is overmodulated.

Spurious emissions — Signals from a transmitter on frequencies other than the operating frequency.

Squelch — Circuitry that mutes the audio output of a receiver when no signal is received. **Carrier squelch** operates only on the presence of a signal carrier. **Tone squelch** requires a specific CTCSS tone to be present before allowing receiver audio to be heard. **Digital Code Squelch** (DCS) requires a continuous sequence of tones.

Squelch tail — The burst of noise heard from an *FM* receiver between when a station stops transmitting and when the receiver's *squelch* circuit mutes the receiver.

Standard offset — The standard transmitter/receiver frequency offset used by a *repeater* on a particular *amateur band*. For example, the standard offset on 2 meters is 600 kHz. Also see *offset frequency*.

Standing-wave ratio (SWR) — A measure of the impedance match between the *transmission line's* characteristic impedance and that of the *load* (usually an *antenna* or *antenna system*). **VSWR** is the ratio of maximum voltage to minimum voltage along the *transmission line* formed by the standing waves that result from power being reflected by the antenna or load. SWR is also the ratio of feed point impedance or load impedance to the feed line's characteristic impedance.

Station license — See *operator/primary station license*.

Sub-audible tone — See *CTCSS*.

Suffix — The letters that follow a *call sign* prefix identifying a specific amateur. Switch — A component used by an operator to connect or disconnect electrical circuits.

SWR meter — A measuring instrument that can indicate when an antenna system is working well. A device used to measure SWR. See *standing wave ratio*.

Tactical call signs — Names used to identify a station's location or function during emergency communications.

Tactical communications — Communications to coordinate actions or logistics during an emergency, disaster, or public service operation.

Telecommand — A one-way radio transmission to start, change or end functions of a device at a distance.

Telemetry — Information about a device sent to a receiving station by radio.

Temporary state of communications emergency — When a disaster disrupts normal communications in a particular area, the FCC can declare this type of emergency. Certain rules may apply for the duration of the emergency.

Ticket — Slang for an amateur radio license.

Time-out timer — A device that limits the amount of time a *repeater* can transmit without a pause by the input signal.

Tolerance — The allowed variation in the dimensions or value of an electrical or mechanical *component*, usually expressed in percent or as a range of values.

Tone scanning — determines what CTCSS access tones are present in specific signal.

Track — To follow a satellite as it travels around the Earth. Tracking software uses the satellite's *Keplerian elements* to determine its location and when it is visible from a specific location.

Traffic — Formal messages exchanged via radio. *Traffic handling* is the process of exchanging traffic. A *traffic net* is a net specially created and managed to handle traffic.

Transformer — An electrical *component* that transfers ac power from one circuit to another by means of a magnetic field shared by two or more *inductors*.

Transient — A short pulse of electrical energy.

Transceiver (XCVR) — A radio transmitter and receiver combined in one unit.

Transmission line — Cable used to connect a transmitter, receiver or transceiver to an antenna or load.

Transmitter (XMTR) — A device that produces radio frequency signals with sufficient power to be useful for communications.

Transponder — A device usually used on satellites that retransmit all signals in a range of frequencies.

Transverter — A device that converts signals so that a transceiver can operate on another band.

Trip — Activate when a threshold is exceeded or an event is detected. A *circuit breaker* trips, opening a circuit, when excessive *current* flow occurs, for example.

Tuned circuit — A circuit with a *resonant frequency* that can be adjusted, usually through the use of adjustable *capacitors* or *inductors*.

Tuning — Adjusting a radio or *circuit* that is frequency-sensitive.

UHF connector — A type of *RF connector* usually used below 500 MHz.

Ultra high frequency (UHF) — The term used for the frequency range from 300 MHz to 3000 MHz (3 GHz).

Ultraviolet (UV) — *Electromagnetic waves* with frequencies greater than visible light. Literally, "above violet," which is the high-frequency end of the visible range.

Unbalanced line — *Transmission line* with one conductor connected to ground, such as coaxial cable.

Uncontrolled environment — Any area in which an RF signal may cause radiation exposure to people who may not be aware of the radiated electric and magnetic fields. The FCC generally considers members of the general public and an amateur's neighbors to be in an uncontrolled RF radiation exposure environment to determine the *maximum permissible exposure* levels. See *controlled environment.*

Unidentified communications or signals — Signals or radio communications in which the transmitting station's *call sign* is not transmitted.

Unintentional radiator — A device that radiates RF signals not required for its normal operation.Universal Licensing System (ULS) — FCC database for all FCC radio services and licensees.

Uplink — Transmitted signals or the range of frequencies for transmissions from Earth to a satellite. See *downlink.*

Upper sideband (USB) — (1) In an AM signal, the *sideband* located above the carrier frequency. (2) The common single-sideband operating mode on the 60, 20, 17, 15, 12 and 10 meter HF amateur bands, and all the VHF and UHF bands.

Vacuum tube — An electronic *component* that operates by controlling *electron* flow between two or more *electrodes* in a vacuum.

Vanity call — A *call sign* selected by the amateur instead of one sequentially assigned by the FCC.

Variable-frequency oscillator (VFO) — An *oscillator* with an adjustable frequency. A VFO is used in receivers and transmitters to control the operating frequency.

Vertical antenna — An antenna with a single vertical radiating *element.* See *ground-plane antenna.*

Very high frequency (VHF) — The term used for the frequency range from 30 MHz to 300 MHz.

Visible horizon — The most distant point one can see visually.

Voice mode (communications) — Any of the several methods used by amateurs to transmit speech that is listened to as it is received. Voice modes include analog modes such as *FM* and SSB as well as digital modes such as D-STAR and DMR.

Voice Over Internet Protocol (VOIP) — A method of sending voice and other audio over the Internet as digital data.

Volt (V) — The basic unit of electric potential or *electromotive force.*

Voltage — A measure of *electric potential* between two points.

Volunteer Examiner (VE) — A licensed amateur who is accredited by a Volunteer Examiner Coordinator (VEC) to administer amateur license examinations.

Volunteer Examiner Coordinator (VEC) — An organization that has entered into an agreement with the FCC to coordinate amateur license examinations.

Voice-Operated Transmission (VOX) — Turning a transmitter on and off under control of the operator's voice.

Watt (W) — The unit of *power* in the metric system.

Waveform — The amplitude of an ac signal over time.

Wavelength —The distance a *radio wave* travels during one *cycle.* The wavelength relates to frequency in that higher frequency waves have shorter wavelengths. Represented by the symbol λ.

Weak-signal — (1) Refers to the use of SSB, digital, or CW on the VHF and UHF bands because they provide better communications at low signal levels than *FM* signals. (2) Any mode of operation that involves very low signal levels.

Whip antenna — An antenna with an *element* made of a single, flexible rod or tube.

Willful interference — Intentional, deliberate obstruction of radio communications.

WSJT — A suite of software programs for weak signal and scatter mode communications that includes the mode FT8.

WWV/WWVH — Radio stations run by the National Institute of Standards and Technology (NIST) to provide accurate time and frequencies.

XCVR — Transceiver

XMTR — Transmitter

Yagi antenna — The most popular type of *directional antenna* or *beam*. It has one *driven element* and one or more *parasitic elements*.

Notes

Notes

Notes

Notes